高等院校技能应用型教材·计算机应用系列

计算机专业英语

（第5版）

卜艳萍　主　编
周　伟　副主编

电子工业出版社
Publishing House of Electronics Industry
北京·BEIJING

内 容 简 介

本书共 5 个单元，第 1 单元介绍计算机硬件知识，内容包括计算机硬件基础、中央处理器、存储器和输入/输出技术。第 2 单元介绍计算机软件知识，内容包括数据结构、系统软件、C 语言与 C++语言、Java 与面向对象程序设计、数据库技术及软件工程。第 3 单元介绍多媒体知识及其应用，内容包括计算机图形学、多媒体、计算机动画、虚拟现实和计算机辅助设计。第 4 单元介绍计算机网络及其应用，内容包括计算机网络基础、无线网络、远程教育、网格计算、电子商务和神经网络与专家系统。第 5 单元介绍计算机安全，内容包括网络安全、防火墙与代理和计算机病毒。

本书适合高等院校、高等职业院校计算机科学与技术、计算机应用及电气工程类相关专业的教学，也可供广大计算机专业技术人员学习和参考。

未经许可，不得以任何方式复制或抄袭本书之部分或全部内容。
版权所有，侵权必究。

图书在版编目（CIP）数据

计算机专业英语/卜艳萍主编. —5 版. —北京：电子工业出版社，2019.7
ISBN 978-7-121-35978-1

Ⅰ．①计… Ⅱ．①卜… Ⅲ．①电子计算机－英语－高等职业教育－教材 Ⅳ．①TP3

中国版本图书馆 CIP 数据核字（2019）第 016675 号

策划编辑：薛华强（xuehq@phei.com.cn）
责任编辑：张　慧　　　　　特约编辑：张燕虹
印　　刷：三河市鑫金马印装有限公司
装　　订：三河市鑫金马印装有限公司
出版发行：电子工业出版社
　　　　　北京市海淀区万寿路 173 信箱　邮编：100036
开　　本：787×1 092　1/16　印张：18.25　字数：607 千字
版　　次：2002 年 6 月第 1 版
　　　　　2019 年 7 月第 5 版
印　　次：2022 年 11 月第 8 次印刷
定　　价：55.00 元

凡所购买电子工业出版社图书有缺损问题，请向购买书店调换。若书店售缺，请与本社发行部联系，联系及邮购电话：（010）88254888，88258888。
质量投诉请发邮件至 zlts@phei.com.cn，盗版侵权举报请发邮件至 dbqq@phei.com.cn。
本书咨询联系方式：（010）88254569，xuehq@phei.com.cn，QQ1140210769。

前 言

根据计算机科学与技术及相关专业的教学特点与要求，本书采用最新的计算机专业技术资料，涵盖计算机硬件和软件知识、多媒体技术、计算机网络知识与安全技术及计算机综合应用等专业知识。"计算机专业英语"是计算机应用类专业学生的重要工具课，是提高学生计算机专业知识和英语综合运用能力的课程。通过该课程的学习，可以提高学生阅读及理解专业英语资料的能力，并掌握计算机专业文章翻译的方法和技巧。

本书共 5 个单元，第 1 单元介绍计算机硬件知识，内容包括计算机硬件基础、中央处理器、存储器和输入/输出技术。第 2 单元介绍计算机软件知识，内容包括数据结构、系统软件、C 语言与 C++语言、Java 与面向对象程序设计、数据库技术及软件工程。第 3 单元介绍多媒体知识及其应用，内容包括计算机图形学、多媒体、计算机动画、虚拟现实和计算机辅助设计。第 4 单元介绍计算机网络及其应用，内容包括计算机网络基础、无线网络、远程教育、网格计算、电子商务和神经网络与专家系统。第 5 单元介绍计算机安全，内容包括网络安全、防火墙与代理和计算机病毒。

本书内容选用计算机各领域的最新知识，考虑与计算机专业课程内容的一致性，并注重实践性和实用性。各单元内容安排合理，课文的词汇量、长度及难度符合学生的学习特点。每个单元配有关键词、注释及大量习题，每个单元后的习题与课文内容配合，以利于学生掌握相关知识。课文的参考译文帮助学生理解课文的内容。

相对于《计算机专业英语》（第 4 版），本书对内容和结构都做了更新。

增加的内容：新增的课文有系统软件、C 语言与 C++语言、Java 与面向对象程序设计、三维图形、网格计算、地理信息系统、神经网络与专家系统、网络安全、防火墙与代理和计算机病毒。第 5 单元新增了语法部分，内容为分词、动名词和动词不定式。

删减的内容：删除了计算机发展、C 语言、操作系统、编译原理、多媒体软件、互联网应用、搜索引擎、电子市场等课文。

修改的内容：按照知识点及相关性将第 4 版的 4 个单元修改为第 5 版的 5 个单元。修改了部分课文的内容及练习题，如 C 语言与 C++语言、计算机图形学、多媒体、计算机网络基础、电子商务、网络安全等。

另外，对附录 A 和附录 B 的计算机专业词汇和缩略词也做了一些更新。

本书由上海交通大学教师卜艳萍担任主编，华东理工大学教师周伟担任副主编。卜艳萍编写了 Unit 1、Unit 2、Unit 5 的 5.3~5.5、附录 A 和附录 B，并完成全书的审校和统稿工作。周伟编写了 Unit 3、Unit 4 和 Unit 5 的 5.1~5.2。赵桂钦、何飞、周烨晴、邱遥等对本书的内容选择和单元结构等提出了有益的建议，在此谨向他们表示衷心的感谢。同时，也对为本书出版而辛勤工作的电子工业出版社的朋友们表示诚挚的谢意。

由于编者水平有限，书中难免有错漏与不当之处，敬请同行与读者批评指正。

主编邮箱：ypbu@sjtu.edu.cn。

编　者

目 录

CONTENTS

Unit 1　Hardware Knowledge ············ 1
　1.1　Hardware Basics ·················· 1
　　1.1.1　Text A ······················· 1
　　1.1.2　Text B ······················· 3
　　1.1.3　Exercises ····················· 5
　1.2　Central Processing Unit ·········· 6
　　1.2.1　Text A ······················· 6
　　1.2.2　Text B ······················· 9
　　1.2.3　Exercises ···················· 11
　1.3　Memory ··························· 12
　　1.3.1　Text A ······················ 12
　　1.3.2　Text B ······················ 15
　　1.3.3　Exercises ···················· 17
　1.4　Input/Output Systems ············ 18
　　1.4.1　Text A ······················ 18
　　1.4.2　Text B ······················ 21
　　1.4.3　Exercises ···················· 23
　1.5　计算机专业英语的词汇特点 ········ 24
　　1.5.1　计算机专业英语的特点 ········ 24
　　1.5.2　专业英语词汇的构成特点 ······ 27
　　1.5.3　专业术语与命令 ·············· 32
　1.6　习题答案与参考译文 ·············· 34
　　1.6.1　第 1 单元习题答案 ············ 34
　　1.6.2　第 1 单元参考译文 ············ 36

Unit 2　Software Knowledge ············ 47
　2.1　Data Structure ·················· 47
　　2.1.1　Text A ······················ 47
　　2.1.2　Text B ······················ 49
　　2.1.3　Exercises ···················· 51
　2.2　System Software ················· 52

　　2.2.1　Text A ······················ 52
　　2.2.2　Text B ······················ 55
　　2.2.3　Exercises ···················· 57
　2.3　C Language and C++ Language ····· 58
　　2.3.1　Text A ······················ 58
　　2.3.2　Text B ······················ 61
　　2.3.3　Exercises ···················· 63
　2.4　Java and Object-Oriented
　　　　Programming ···················· 64
　　2.4.1　Text A ······················ 64
　　2.4.2　Text B ······················ 67
　　2.4.3　Exercises ···················· 69
　2.5　Database Technologies ············ 70
　　2.5.1　Text A ······················ 70
　　2.5.2　Text B ······················ 73
　　2.5.3　Exercises ···················· 74
　2.6　Software Engineering ············· 75
　　2.6.1　Text A ······················ 75
　　2.6.2　Text B ······················ 78
　　2.6.3　Exercises ···················· 80
　2.7　计算机专业英语的阅读与翻译 ······ 81
　　2.7.1　阅读方法 ···················· 81
　　2.7.2　翻译的方法与过程 ············ 83
　2.8　习题答案与参考译文 ·············· 91
　　2.8.1　第 2 单元习题答案 ············ 91
　　2.8.2　第 2 单元参考译文 ············ 95

Unit 3　Multimedia and Its
　　　　Applications ···················· 109
　3.1　Computer Graphics ··············· 109
　　3.1.1　Text A ····················· 109
　　3.1.2　Text B ····················· 112

	3.1.3	Exercises ········· 114
3.2	Multimedia ············ 115	
	3.2.1	Text A ············ 115
	3.2.2	Text B ············ 119
	3.2.3	Exercises ········· 120
3.3	Computer Animation ····· 122	
	3.3.1	Text A ············ 122
	3.3.2	Text B ············ 124
	3.3.3	Exercises ········· 127
3.4	Virtual Reality ··········· 128	
	3.4.1	Text A ············ 128
	3.4.2	Text B ············ 131
	3.4.3	Exercises ········· 133
3.5	Computer Aided Design ·· 135	
	3.5.1	Text A ············ 135
	3.5.2	Text B ············ 137
	3.5.3	Exercises ········· 139
3.6	专业英语中被动语态与 长句的运用 ············ 140	
	3.6.1	被动语态的运用 ····· 140
	3.6.2	长句的运用 ········ 144
3.7	习题答案与参考译文 ···· 147	
	3.7.1	第3单元习题答案 ··· 147
	3.7.2	第3单元参考译文 ··· 150

Unit 4　Computer Network and Its Applications ············ 162

4.1	Computer Network Basics ···· 162	
	4.1.1	Text A ············ 162
	4.1.2	Text B ············ 165
	4.1.3	Exercises ········· 166
4.2	Wireless Network ······· 168	
	4.2.1	Text A ············ 168
	4.2.2	Text B ············ 171
	4.2.3	Exercises ········· 173
4.3	Distance Education ······ 174	
	4.3.1	Text A ············ 174
	4.3.2	Text B ············ 177
	4.3.3	Exercises ········· 179
4.4	Grid Computing ········ 180	
	4.4.1	Text A ············ 180
	4.4.2	Text B ············ 183
	4.4.3	Exercises ········· 185
4.5	Electronic-Commerce ··· 186	
	4.5.1	Text A ············ 186
	4.5.2	Text B ············ 190
	4.5.3	Exercises ········· 192
4.6	Neural Networks and Expert System ················ 193	
	4.6.1	Text A ············ 193
	4.6.2	Text B ············ 196
	4.6.3	Exercises ········· 198
4.7	定语从句与状语从句 ···· 200	
	4.7.1	定语从句 ·········· 200
	4.7.2	状语从句 ·········· 202
4.8	习题答案与参考译文 ···· 206	
	4.8.1	第4单元习题答案 ··· 206
	4.8.2	第4单元参考译文 ··· 210

Unit 5　Computer Security ············ 225

5.1	Network Security ······· 225	
	5.1.1	Text A ············ 225
	5.1.2	Text B ············ 228
	5.1.3	Exercises ········· 230
5.2	Firewalls and Proxies ···· 231	
	5.2.1	Text A ············ 231
	5.2.2	Text B ············ 234
	5.2.3	Exercises ········· 236
5.3	Computer Virus ········· 238	
	5.3.1	Text A ············ 238
	5.3.2	Text B ············ 241
	5.3.3	Exercises ········· 243
5.4	分词、动名词和动词不定式 ··· 245	
	5.4.1	分词 ·············· 245
	5.4.2	动名词 ············ 246
	5.4.3	动词不定式 ········ 247
5.5	习题答案与参考译文 ···· 249	
	5.5.1	第5单元习题答案 ··· 249
	5.5.2	第5单元参考译文 ··· 251

附录A　计算机专业英语词汇表 ············ 260

附录B　计算机专业英语缩略词 ············ 275

参考文献 ············ 286

Unit 1　Hardware Knowledge

1.1　Hardware Basics

1.1.1　Text A

Combinational Circuit

A combinational circuit is a connected arrangement of logic gates with a set of inputs and outputs. At any given time, the binary values of the outputs are a function of the binary combination of the inputs. A block diagram of a combinational circuit is shown in Fig. 1-1. The n binary input variables come from an external source, the m binary output variables go to an external destination, and in between there is an interconnection of logic gates. A combinational circuit transforms binary information from the given input data to the required output data. [1]

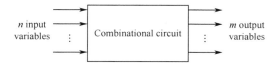

Fig. 1-1　A block diagram of a combinational circuit

A combinational circuit can be described by a truth table showing the binary relationship between the n input variables and the m output variables. The truth table lists the corresponding output binary values for each of 2^n input combination. A combinational circuit can also be specified with m Boolean functions, one for each output variable. [2] Each output function is expressed in terms of the n input variables.

Logic Systems

In a DC, or level-logic system, a bit is implemented as one of two voltage levels. If the more positive voltage is the 1 level and the other is the 0 level, the system is said to employ DC positive logic. [3] On the other hand, a DC negative-logic system is one which designates the more negative voltage state of the bit as the 1 level and the more positive as the 0 level. It should be emphasized that the absolute values of the two voltages are of no significance in these definitions. In particular, the 0 state need not represent a zero voltage level (although in some systems it might).

The parameters of a physical device are not identical from sample to sample, and they also vary with temperature. Furthermore, ripple or voltage spikes may exist in the power supply or ground leads, and other sources of unwanted signals, called noise, may present in the circuit. For these reasons, the digital levels are not specified precisely, but each state is defined by a voltage range about a designated level, such as 4V±1V and 0.2V±0.2V.

In a dynamic, or pulse-logic system, a bit is recognized by the presence or absence of a pulse. A 1 signifies the existence of a positive pulse in a dynamic positive-logic system; a negative pulse denotes a 1 in a dynamic negative-logic system. In either system a 0 at a particular input (or output) at a given instant of time designates that no pulse is present at that particular moment.

Logic Circuits

The design of digital computers is based on a logical methodology called Boolean Algebra which uses three basic operations: logical addition, called the OR function; logical multiplication, called the AND function; and logical complementation, called the NOT function. The variables in Boolean algebra are binary, namely, the resulting variable of an operation or a set of operations can have only one of the two values: One or Zero. These two values may also be interpreted as being True or False, Yes or No, and Positive or Negative.

A switch is ideally suited to represent the value of any two-state variable because it can only be "off" or "on". [4]

There are only three basic logic operations: the conjunction (logical product) commonly called AND; the disjunction (logic sum) commonly called OR; and the negation commonly called NOT.

Key Words

arrangement	排列，整理
combinational	组合的，合并的
complementation	互补，补足
conjunction	结合，联合，联系
designate	指出，指明
diagram	图形，图解
disjunction	分离，析取
interconnection	互连网络
methodology	方法学，方法论
multiplication	倍增，乘法运算
negation	否定，拒绝
negative-logic	负逻辑
represent	表示，代表
ripple	脉动，波动

significance　　　　　　　　重要性，意义
spike　　　　　　　　　　　尖峰信号

Notes

[1] A combinational circuit transforms binary information from the given input data to the required output data.

说明：本句的介词短语"from…to…"是宾语补足语，进一步说明宾语"binary information"。

译文：组合电路通过传输二进制信息，使得给定的输入数据产生了所需要的输出数据。

[2] A combinational circuit can also be specified with m Boolean functions, one for each output variable.

说明：本句的"one for each output variable"是同位语。

译文：组合电路也能规定 m 种布尔函数，每种函数对应一个输出变量。

[3] If the more positive voltage is the 1 level and the other is the 0 level, the system is said to employ DC positive logic.

说明：本句的"If the more positive voltage…"是条件状语。"the system is said to…"用被动语态表示客观叙述。

译文：如果较高的电压为 1 电平，而另一个是 0 电平，则称该系统使用的是直流正逻辑。

[4] A switch is ideally suited to represent the value of any two-state variable because it can only be "off" or "on".

说明：本句的 because it can only be "off" or "on"是原因状语从句。

译文：因为开关只能是"关"或"开"两个状态，所以它最适合表示两个状态的变量值。

1.1.2　Text B

Bill Gates is an American business magnate, computer programmer and philanthropist. Gates is the former chief executive officer and current chairman of Microsoft, the world's largest personal-computer software company, which he co-founded with Paul Allen. He is consistently ranked among the world's wealthiest people and was the wealthiest overall from 1995 to 2007. During his career at Microsoft, Gates held the positions of CEO and chief software architect, and remains the largest individual shareholder, with 8 percent of the common stock. He has also authored or co-authored several books.

Gates is one of the best-known entrepreneurs of the personal computer revolution. In the later stages of his career, Gates has pursued a number of philanthropic endeavors, donating large amounts of money to various charitable organizations and scientific research programs through the Bill & Melinda Gates Foundation, established in 2000.

Gates was born in Seattle, Washington. His father was a prominent lawyer, and his mother served on the board of directors for First Interstate BancSystem. At 13 he enrolled in the Lakeside School, an exclusive preparatory school. Gates graduated from Lakeside School in

1973. He scored 1590 out of 1600 on the SAT and enrolled at Harvard College in the autumn of 1973. While at Harvard, he met Steve Ballmer, who later succeeded Gates as CEO of Microsoft.

In his sophomore year, Gates devised an algorithm for pancake sorting as a solution to one of a series of unsolved problems presented in a combinatorics class by Harry Lewis, one of his professors. Gates's solution held the record as the fastest version for over thirty years; its successor is faster by only one percent. His solution was later formalized in a published paper in collaboration with Harvard computer scientist Christos Papadimitriou.

Pancake sorting is a variation of the sorting problem in which the only allowed operation is to reverse the elements of some prefix of the sequence. Unlike a traditional sorting algorithm, which attempts to sort with the fewest comparisons possible, the goal is to sort the sequence in as few reversals as possible. This operation can be visualized by thinking of a stack of pancakes in which one is allowed to take the top k pancakes and flip them. A variant of the problem is concerned with burnt pancakes, where each pancake has a burnt side and all pancakes must, in addition, end up with the burnt side on top.

Gates did not have a definite study plan while a student at Harvard and spent a lot of time using the school's computers. Gates remained in contact with Paul Allen, and he joined him at Honeywell during the summer of 1974. The following year saw the release of the MITS Altair 8800 based on the Intel 8080 CPU, and Gates and Allen saw this as the opportunity to start their own computer software company. Gates dropped out of Harvard at this time. He had talked this decision over with his parents, who were supportive of him after seeing how much Gates wanted to start a company.

After reading the January 1975 issue of Popular Electronics that demonstrated the Altair 8800, Gates contacted Micro Instrumentation and Telemetry Systems (MITS), the creators of the new microcomputer, to inform them that he and others were working on a BASIC interpreter for the platform. In reality, Gates and Allen did not have an Altair and had not written code for it; they merely wanted to gauge MITS's interest. MITS president Ed Roberts agreed to meet them for a demo, and over the course of a few weeks they developed an Altair emulator that ran on a minicomputer, and then the BASIC interpreter.

The demonstration, held at MITS's offices in Albuquerque was a success and resulted in a deal with MITS to distribute the interpreter as Altair BASIC. Paul Allen was hired into MITS, and Gates took a leave of absence from Harvard to work with Allen at MITS in Albuquerque in November 1975. They named their partnership "Micro-Soft" and had their first office located in Albuquerque. Within a year, the hyphen was dropped, and on November 26, 1976, the trade name "Microsoft" was registered with the Office of the Secretary of the State of New Mexico. Gates never returned to Harvard to complete his studies.

Gates stepped down as chief executive officer of Microsoft in January 2000. He remained as chairman and created the position of chief software architect. In June 2006, Gates announced that he would be transitioning from full-time work at Microsoft to part-time work, and full-time work at the Bill & Melinda Gates Foundation.

Key Words

charitable	仁爱的，慈善的
endeavor	努力，尽力
entrepreneur	企业家，创业人
foundation	基金，基础，建设
hyphen	连字号，连字符
interpreter	解释器，翻译程序
pancake	薄煎饼，烙饼
philanthropist	慈善家
preparatory	预备的，准备的，筹备的
release	发布，放开
shareholder	股东
wealthiest	富人

1.1.3 Exercises

1. Translate the following phrases into English

(1)逻辑电路
(2)输出变量
(3)二进制信息
(4)组合电路
(5)正向脉冲

2. Translate the following phrases into Chinese

(1)Boolean Algebra
(2)chief executive officer
(3)full-time work
(4)Harvard College
(5)scientific research program

3. Identify the following to be True or False according to the text

(1)In a DC, or level-logic system, a bit is implemented as one of two voltage levels.
(2)Gates stepped down as chief executive officer of Microsoft in January 2010.
(3)In a pulse-logic system a bit is not recognized by the presence or absence of a pulse.
(4)The output values are a function of the combination of the inputs in a combination circuit.
(5)Gates had a definite study plan while a student at Harvard.

4. Reading Comprehension

(1) The parameters of a physical device are not identical from sample to sample, and they also vary with _____.

    a. height

    b. weight

    c. temperature

    d. length

(2) Unlike a traditional sorting _____, which attempts to sort with the fewest comparisons possible, the goal is to sort the sequence in as few reversals as possible.

    a. algorithm

    b. data

    c. sequence

    d. computer

(3) A combinational circuit transforms _____ information from the given input data to the required output data.

    a. hexadecimal

    b. binary

    c. octal

    d. decimal

(4) The _____ lists the corresponding output binary values for each of 2^n input combination.

    a. paper

    b. truth list

    c. table

    d. truth table

1.2 Central Processing Unit

1.2.1 Text A

The basic job of computers is the processing of information. For this reason, computers can be defined as devices which accept information in the form of instructions called a program and characters called data, perform mathematical and logical operations on the information, and then supply results of these operations. The program, which tells the computers what to do and the data, which provide the information needed to solve the problem, are kept inside the computer in a place called memory.[1]

Computers are thought to have many remarkable powers. However, most computers, whether large or small, have three basic capabilities.

First, computers have circuits for performing arithmetic operations, such as addition, subtraction, division, multiplication and exponentiation.

Second, computers have a means of communicating with the user. After all, if we couldn't feed information in and get results back, these machines would not be of much use.

Third, computers have circuits which can make decisions. The kinds of decisions which computer circuits can make are of the type: Is one number less than another? Are two numbers equal? And, is one number greater than another?

A CPU can be a single microprocessor chip, a set of chips, or a box of boards of transistors, chips, wires, and connectors. Differences in CPUs distinguish mainframes, mini-computers and micro-computers. A processor is composed of two functional units: a control unit and an arithmetic/logic unit, and a set of special workspaces called registers.

The control unit

The control unit is the functional unit that is responsible for supervising the operation of the entire computer system. In some ways, it is analogous to a telephone switchboard with intelligence because it makes the connections between various functional units of the computer system and calls into operation each unit that is required by the program currently in operation. The control unit fetches instructions from memory and determines their type or decodes them. It then breaks each instruction into a series of simple small steps or actions. By doing this, it controls the step-by-step operation of the entire computer system.

The Arithmetic/Logic Unit

The Arithmetic/Logic Unit (ALU) is the functional unit that provides the computer with logical and computational capabilities.[2] Data are brought into the ALU by the control unit, and the ALU performs whatever arithmetic or logic operations are required to help carry out the instructions. Arithmetic operations include adding, subtracting, multiplying, and dividing. Logic operations make a comparison and take action based on the results. For example, two numbers might be compared to determine if they are not equal. If they are equal, processing will continue; if they are not equal, processing will stop.

Registers

A register is a storage location inside the processor. Registers in the control unit are used to keep track of the overall status of the program that is running. Control unit registers store information such as the current instruction, the location of the next instruction to be executed, and the operands of the instruction. In the ALU, registers store data items that are added, subtracted, multiplied, divided, and compared. Other registers store the results of arithmetic and logic operations.

Instruction

An instruction is made up of operations that specify the function to be performed and operands that represent the data to be operated on. For example, if an instruction is to perform the operation of adding two numbers, it must know what the two numbers are and where the two numbers are. [3] When the numbers are stored in the computer's memory, they have an address to indicate where they are, so if an operand refers to data in the computer's memory it is called an address. The processor's job is to retrieve instructions and operands from memory and to perform each operation. Having done that, it signals memory to send it the next instruction.

The CPU executes each instruction in a series of small steps:

1. Fetch the next instruction from memory into the instruction register.
2. Change the program counter to point to the following instruction.
3. Determine the type of instruction just fetched.
4. If the instruction uses data in memory, determine where they are.
5. Fetch the data into internal CPU registers.
6. Execute the instruction.
7. Store the results in the proper place.

Go to step 1 to begin executing the following instruction.

This sequence of steps is frequently referred to as the fetch-decode-execute cycle. It is central to the operation of all computers. This step-by-step operation is repeated over and again at awesome speed. A timer called a clock releases precisely timed electrical signals that provide a regular pulse for the processor's work. [4] The term that is used to measure the computer's speed is borrowed from the domain of electrical engineering and is called a megahertz (MHz) which means million cycles per second.

Key Words

address	地址，寻址
analogous	类似的，相似的，可比拟的
arithmetic	算术的
awesome	惊人的，令人敬畏的
capability	性能，能力
distinguish	区别，辨别
exponentiation	幂运算
fetch	获取，取得
instruction	指令
intelligence	智能，智慧，智力
mainframe	大型机
manipulation	操作，处理
microelectronic	微电子的

operand	操作数
remarkable	显著的，不平常的
sequence	顺序，序列
supervise	监督，管理，指导
switchboard	配电盘，接线总机
workspace	工作空间，工作区

Notes

[1] The program, which tells the computers what to do and the data, which provide the information needed to solve the problem, are kept inside the computer in a place called memory.

这里的主语是"the program and the data"，由 which 引导的两个定语从句分别修饰 the program 和 the data。

译文：程序的作用是指示计算机如何工作，而数据则为解决问题提供所需要的信息，两者都存储在存储器里。

[2] The Arithmetic/Logic Unit (ALU) is the functional unit that provides the computer with logical and computational capabilities.

本句由"that"引导定语从句，修饰"the functional unit"。

译文：算术/逻辑单元（ALU）是为计算机提供逻辑及计算能力的功能部件。

[3] For example, if an instruction is to perform the operation of adding two numbers, it must know what the two numbers are and where the two numbers are.

这里的"what the two numbers are and where the two numbers are"是宾语，它由两个并列的从句组成。

译文：例如，如果一条指令要完成两数相加的操作，它就必须知道：这两个数是什么？这两个数在哪里？

[4] A timer called a clock releases precisely timed electrical signals that provide a regular pulse for the processor's work.

本句中的"that provide a regular pulse for the processor's work"修饰 electrical signals。

译文：一个称为时钟的计时器准确地发出定时电信号，该信号为处理器工作提供有规律的脉冲信号。

1.2.2 Text B

The CPU in a microcomputer is actually one relatively small integrated circuit or chip. Although most CPU chips are smaller than a lens of a pair of glasses, the electronic components they contain would have filled a room a few decades ago. Using advanced microelectronic techniques, manufacturers can cram tens of thousands of circuits into tiny layered silicon chips that work dependably and use less power. The CPU coordinates all the activities of the various components of the computer. It determines which operations should be carried out and in what order. The CPU can also retrieve information from memory and can store the results of manipulations back into the memory unit for later reference.

Microcomputer, or micro for short, is a kind of computers. It was born in the early 1970s. The computer's brain is called the microprocessor. That's the main chip in a computer that does all the work. It's also the center of activity on the motherboard. It interprets and executes the instructions which comprise a computer program. The CPU consists of an arithmetic unit and its associated circuitry, known as the arithmetic and logic unit, and an instruction counter and decoder. The CPU can perform only one operation at a time. Essentially, numerically coded instructions are stored in the computer's high-speed storage, or primary storage. The CPU takes the instructions one at a time and executes them. The numerical coding of the instruction tells the CPU which operation to perform and where the data upon which the operation is to take place is stored.

The central processor of the micro, called the microprocessor, is built as a single semiconductor device; that is, the thousands of individual circuit elements necessary to perform all the logical and arithmetic functions of a computer are manufactured as a single chip. A complete microcomputer system is composed of a microprocessor, a memory and some peripheral equipment. The processor, memory and electronic controls for the peripheral equipment are usually put together on a single or on a few printed circuit boards. Systems using microprocessors can be hooked up together to do the works that until recently only minicomputer systems were capable of doing. Micros generally have somewhat simpler and less flexible instruction sets than minis, and are typically much slower. Similarly, minis are available with much larger primary memory sizes. Micros are becoming more powerful and converging with minicomputer technology.

The microprocessor is essentially a small calculator. It does basic calculator like things—adding, subtracting, multiplying, and dividing values stored in the computer's memory. Computer programs tell the microprocessor what to do, which is how everything works inside a PC.

Other terms for the microprocessor include the processor; the central processing unit (CPU); and the number of the microprocessor, such as 8088, 80286, 80386,80486, and so on . There are three main varieties of microprocessors for PCs: the 8088/8086; the 80286, or AT microprocessor; and the 386 family of microprocessor. There is no 586 microprocessor. Instead of calling it a number, the company that manufactured it (Intel) called it the Pentium. It's the micro part of microprocessor that led old-time computer users to call PCs microprocessor. This may have applied to the first microprocessor, but today's powerhouse PCs are anything but micro.

Key Words

  calculator   计算器，计算者
  circuitry   电路学，电路系统
  dependably   可信任地

essentially	本质上，基本上
flexible	灵活的，易适应的
numerically	数字化地，用数表示地
peripheral	周围的，外围的
powerhouse	发电所，动力室
reference	参考，参照
relatively	相对地，比较地
somewhat	有点，有些，稍微

1.2.3　Exercises

1. Translate the following phrases into English

(1)取指—译码—执行

(2)算术/逻辑运算

(3)微电子技术

(4)印制电路板

(5)外围设备

2. Translate the following phrases into Chinese

(1)current instruction

(2)instruction register

(3)program counter

(4)retrieve information

(5)high-speed storage

3. Identify the following to be True or False according to the text

(1)In the ALU, registers store data items that are added, subtracted, multiplied, divided, and compared.

(2)Registers in the control unit are used to keep track of the overall status of the program.

(3)In the ALU, registers only store the results of arithmetic and logic operations.

(4)Microcomputer was born in the early 1970s.

(5)To store the results in the proper place is done by ALU.

4. Reading Comprehension

(1)A processor is composed of two functional units, they are＿＿＿＿＿＿＿＿＿＿.

a. an arithmetic/logic unit and a storage unit

b. a control unit and some registers

c. a control unit and an arithmetic/logic unit

d. some registers and an arithmetic/logic unit

(2)The control unit fetches ＿＿＿＿＿＿＿＿＿＿ from memory and decodes them.

a. data

b. information

c. results

d. instructions

(3)_____ is a storage location inside the processor.

a. A register

b. ALU

c. Control unit

d. Memory

(4)The CPU executes each instruction in a series of steps, the sequence is_____.

a. execute-fetch-decode

b. fetch-decode-execute

c. decode-execute-fetch

d. fetch-execute-storage

1.3 Memory

1.3.1 Text A

A memory cell is a circuit, or in some cases just a single device, that can store a bit of information. A systematic arrangement of memory cells constitutes a memory. The memory must also include peripheral circuits to address and write data into the cells as well as detect data that are stored in the cells.

Two basic types of semiconductor memory are considered. The first is the Random Access Memory (RAM), a read-write memory, in which each individual cell can be addressed at any particular time. The access time to each cell is virtually the same. Implicit in the definition of the RAM is that both the read and the write operations are permissible in each cell with also approximately the same access time.

A second class of semiconductor memory is the Read-Only Memory (ROM). The set of data in this type of memory is generally considered to be fixed, although in some designs the data can be altered.[1] However, the time required to write new data is considerably longer than the read access time of the memory cell. A ROM may be used, for example, to store the instructions of a system operating program.

A volatile memory is one that loses its data when power is removed from the circuit, while non-volatile memory retains its data even when power is removed. In general, a Random Access Memory is a volatile memory, while Read-Only Memories are nonvolatile.

Two type of RAM are the static RAM (SRAM) and dynamic RAM (DRAM). A static RAM consists of a basic bi-stable flip-flop circuit that needs only a DC current or voltage applied to retain its memory. Two stable states exist, defined as logic 1 and logic 0. A dynamic RAM is an

MOS memory that stores one bit of information as charge on a capacitor. Since the charge on the capacitor delays with a finite time constant (milliseconds), a periodic refresh is needed to restore the charge so that the dynamic RAM does not lose its memory.

The advantage of the SRAM is that this circuit does not need the additional complexity of a refresh cycle and refresh circuitry, but the disadvantage is that this circuit is fairly large. In general, a SRAM requires six transistors. The advantage of a DRAM is that it consists only one transistor and one capacitor, but the disadvantage is the required refresh circuitry and refresh cycles.

There are two general types of ROM. The first is programmed either by the manufacturer (mask programmable) or by the user (programmable, or PROM). Once the ROM has been programmed by either method, the data in the memory are fixed and cannot be altered. The second type of ROM may be referred to as an alterable ROM in that the data in the ROM may be reprogrammed if desired. This type of ROM may be called an EPROM (erasable programmable ROM), EEPROM (electrically erasable PROM), or flash memory. As mentioned, the data in these memories can be reprogrammed although the time involved is much longer than the read access time. In some cases, the memory chip may actually have to be removed from the circuit during the reprogramming process.

The basic memory architecture has the configuration shown in Figure 1-2. The terminal connections may include inputs, outputs, addresses, and read and write controls. The main potion of the memory involves the data storage. A RAM memory will have all of the terminal connections mentioned, whereas a ROM memory will not have the inputs and the write controls.

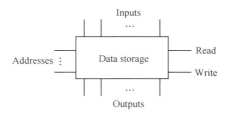

Fig. 1-2 Basic memory architecture

Computer memory is measured in kilobytes or megabytes of information. A byte is the amount of storage needed to hold one character, such as a letter or a numeric digit. One kilobyte (KB) equals 1024 bytes, and one megabyte (MB) is about 1 million bytes. Software requires the correct amount of RAM to work properly. If you want to add new software to your computer, you can usually find the exact memory requirements on the software packaging.[2]

Memories consist of a number of cells, each of which can store a piece of information. Each cell has a number, called its address, by which programs can refer to it. If a memory has n cells, they will have addresses 0 to $n-1$. All cells in a memory contain the same number of bits. If a cell consists of k bits, it can hold any one of 2^k different bit combinations. Note that adjacent cells have consecutive addresses.

Computers that use the binary number system also express memory addresses as binary numbers. If an address has m bits, the maximum number of cells directly addressable is 2^m. The number of bits in the address is related to the maximum number of directly addressable cells in the memory and is independent of the number of bits per cell. [3] A memory with 2^{12} cells of 8 bits each and a memory with 2^{12} cells of 64 bits each would each need 12-bit addresses.

The significance of the cell is that it is the smallest addressable unit. In recent years, most computer manufactures have standardized on an 8-bit cell, which is called a byte. Bytes are grouped into words. A computer with a 16-bit word has 2 bytes/word, whereas a computer with a 32-bit word has 4 bytes/word. The significance of a word is that most instructions operate on entire words, for example, adding two words together. Thus a 32-bit machine will have 32-bit registers and instructions for moving, adding, subtracting, and otherwise manipulating 32-bit words. [4]

Key Words

adjacent	邻近的，接近的
alterable	可改变的，可修改的
approximately	近似地，大约地
capacitor	电容
considerably	相当地，非常地
dynamic	动态的，动力的
erasable	可消除的，可擦拭的，可删除的
kilobyte	千字节（KB）
manufacturer	厂商，制造业者
megabyte	兆字节（MB）
millisecond	毫秒
nonvolatile	非易失性的
programmable	可编程的
refresh	刷新，更新
stable	稳定的
systematic	有系统的，成体系的
volatile	易失去的，易变的

Notes

[1] The set of data in this type of memory is generally considered to be fixed, although in some designs the data can be altered.

说明：本句的主句用被动语态表示客观事实，"although"引导的是让步状语从句。

译文：尽管这类存储器中所设置的数据在某些设计中可以改变，但这些数据通常是固定的。

[2] If you want to add new software to your computer, you can usually find the exact

memory requirements on the software packaging.

说明：本句由"if"引导条件状语从句，"on the software packaging"是状语。

译文：如要给计算机增加新的软件，在软件包装上通常可以找到该软件所需要的确切内存容量。

[3] The number of bits in the address is related to the maximum number of directly addressable cells in the memory and is independent of the number of bits per cell.

说明：本句中，of directly addressable cells in the memory 修饰 the maximum number。

译文：地址的位数与存储器可直接寻址的最大单元数量有关，而与每个单元的位数无关。

[4] Thus 32-bit machine will have 32-bit registers and instructions for moving, adding, subtracting, and otherwise manipulating 32-bit words.

说明：本句的"for moving, adding…"是宾语补足语，进一步说明宾语"registers and instructions"的功能。

译文：因而 32 位机器则有 32 位的寄存器和指令，以实现传送、加法、减法和其他 32 位字的操作。

1.3.2　Text B

In the 1970s, there was a further development which revolutionized the computer field. This was the ability to etch thousands of integrated circuits onto a tiny piece (chip) of silicon, which is a non-metallic element with semiconductor characteristics. Chips have thousands of identical circuits, each one capable of storing one bit. Because of the very small size of the chip, and consequently of the circuits etched on it, electrical signals do not have to travel far; hence, they are transmitted faster. Moreover, the size of the components containing the circuitry can be considerably reduced, a step which has led to the introduction of both minis and micros. As a result, computers have become smaller, faster, and cheaper. There is one problem with semiconductor memory, however, when power is removed, information in the memory is lost, unlike core memory, which is capable of retaining information during a power failure.

The 80x86 processors, operating in real mode, have physical address-ability to 1 megabyte of memory. EMS was developed to allow real mode processing to have access to additional memory. It uses a technique called paging, or bank switching. The requirements for expanded memory include additional hardware and a software device driver. The bank switching registers act as gateways between the physical window within the 1 megabyte space and the logical memory that resides on the expanded memory board. The device driver, called the expanded memory manager (EMM), controls the registers so that a program's memory accesses can be redirected throughout the entire of available expanded memory.

To access expanded memory, a program needs to communicate with the EMM. Communication with the EMM is similar to making calls to DOS. The program sets up the proper CPU registers and makes a software interrupt request. More than 30 major functions are defined, and applications and operating systems are given control over expanded memory. When a program allocates expanded memory pages, the EMM returns a handle to the requesting program. This handle is then used in future calls to the EMM to identify which block of logical pages is being manipulated.

A variety of different types of cache (disk cache, memory cache, processor cache) can improve overall system performance. Although most high-level systems include cache in system design, a cache can be optionally implemented on almost any system—from a low-level 8086 system on up to the highest performance i486-based system.

In the case of hard disk cache, there are two general approaches to cache implementation. The two approaches primarily differ in terms of where the memory in the cache resides. The first and most commonly implemented form of disk cache uses extended memory. In a microcomputer, the extended memory of 1 megabyte or larger can be assigned as disk cache memory. The higher the percentage of "hits" (calls to the disk that can be read from cache, rather than from the disk) , the greater the overall performance of the system. For word processing or manipulation of small files, a huge cache may be overkill. For manipulation of large database files of complex graphics, a cache using extended memory can provide a significant performance increase. The second form of disk cache is performed by cache controllers. These controllers not only control the read/write operations of the attached hard disk drives, they also provide a cache that performs many of the functions of the system-based extended memory caches. However, by placing the cache on the controller, performance of a cache can be enhanced over that of most extended memory caches.

High-speed memory is used on cache controllers. When the system makes a call to the hard disk, the controller in cache determines whether or not the data that's being called for is in the cache or on the disk. When using an extended memory cache, the cache intercepts calls for reads from the hard disk and checks to see if the data called for by the system resides in the cache. If the data is not in the cache, a read instruction is sent to the disk controller. This process takes time. Not only is the extended memory used for such caches usually slower than that used in cache controllers, an extra step is involved before the disk controller can be instructed to go to the disk to retrieve the data it seeks. However, by allowing the processor to make calls to the cache controller, which then determines whether or not the disk actually must be read, performance can be significantly enhanced. With intelligent cache controllers, significant performance improvements have been claimed by a number of cache controller manufacturers. Cache controllers have been used on mainframe computer systems for many years; their migration to the micro could be seen as a logical next step.

Key Words

communicate	通信，传递
etch	蚀刻
expanded	扩展的，扩充的，延伸的
failure	失败，缺乏，不足

gateway	网关，信闸，入口处
migration	迁移，移植，移动
optionally	随意地
percentage	百分率，百分比
throughout	遍及，贯穿，在所有方面

1.3.3 Exercises

1. Translate the following phrases into English

(1)易失性存储器

(2)实模式

(3)寻址能力

(4)闪存

(5)刷新电路

2. Translate the following phrases into Chinese

(1)software interrupt

(2)expanded memory

(3)refresh cycle

(4)Read Only Memory

(5)Random Access Memory

3. Identify the following to be True or False according to the text

(1)Both static and dynamic RAM cells are read-write memory.

(2)Nonvolatile memory loses its data when power is removed from the circuit.

(3)ROM does not have the inputs and the write controls.

(4)The memory addresses are expressed as binary numbers.

(5)RAM can be used to store the instructions of a system program.

4. Reading Comprehension

(1)One megabyte equals approximately _____.

a. 1,048,576 bytes

b. 1024 bytes

c. 65,535 bytes

d. 10,000 bytes

(2)If a cell consist of n bits, it can hold any one of _____.

a. $2n$ different bit combinations

b. 2^{n-1} different bit combinations

c. 2^n different bit combinations

d. n different bit combinations

(3)When power is removed, information in the semiconductor memory is _____.

a. reliable

b. lost

c. manipulated

d. remain

(4) A periodic refresh is needed to restore the information for the _____.

a. SRAM

b. EPROM

c. DRAM

d. EEPROM

1.4　Input/Output Systems

1.4.1　Text A

Keyboard

If you are familiar with a typewriter, you'll find the layout of the computer keyboard very similar. You can use your keyboard for many purposes:

- Typing information.
- Entering numbers with the numeric keypad.
- Requesting specific functions.
- Performing system functions with key combinations.
- Moving around the computer screen.

The keyboard has letter keys, punctuation keys, and a spacebar. It also has functions, numeric, and arrow keys. How you use the keys depends on the software installed on your computer. The documentation that comes with your software has information about specific key functions. You will probably notice a difference between the touch (response) on a computer keyboard and the response of a typewriter. A computer keyboard is so responsive that you can type using a light touch. When you hold down a character key, the character continues to type. This is called the typematic effect of a computer keyboard.

Monitor

Monitors maybe are one of the most important output devices. Computers only use monitors to show you exciting operation results or marvelous and vivid pictures. Monitors also are the best windows for conversation between users and computers. So, many users select monitors carefully. Which parameters or indexes ought be paid attention to when you select a monitor? We provide some here for your reference.

Element Distance: The distance between two picture elements in horizontal direction is called element distance here and its current value in most PC monitors is 0.28mm. If the value is smaller, the screen is more distinct.

Video Bandwidth: It is an important concept in monitor technology. It is related to the highest work frequency of the monitor. It is from tens MHz to hundreds MHz.

Solution: It is another important parameter of a monitor. It's higher, the view on a screen is clearer. Solution means the sum of all picture elements on a screen.

Scan Style: The scan style of a electron gun in a tube is divided into two styles: interlace and non-interlace. In interlace style, electron-beam sweeps elements in odd rows first time and does elements in even rows second time. [1] A frame to be renewed needs sweeping two times. In non-interlace style, electron-beam sweeps all elements only in one time. In non-interlace work style, the monitor works better and gives clear pictures without flash.

Mouse

The interface between a mouse and a system can take one of two forms: the mouse either generates a series of pulses when it is moved (using the LED and detector to generate the pulses), or it increments and decrements counters. The processor can periodically read these counters, or count up the pulses, and determine how far the mouse has moved since it was last examined. The system then moves the cursor on the screen appropriately. This motion appears smooth because the rate at which you can move the mouse is slow compared with the rate at which the processor can read the mouse status and move the cursor on the screen.

Most mice also include one or more buttons, and the system must be able to detect when a button is depressed. By monitoring the status of the button, the system can also differentiate between clicking the button and holding it down. [2] Of course, the mapping between the counters and the button position and what happens on the screen is totally controlled by software. That's why, for example, the rate at which the mouse moves across the screen and the rate at which single and double clicks are recognized can usually be set by the user. Similarly, software interpretation of the mouse position means that the cursor doesn't jump completely off the screen when the mouse is moved a long distance in one direction. [3]

Optical Disks

An optical disk is a disk on which data are encoded for retrieval by a laser. Optical disks offer information densities far beyond the range of current magnetic mass-storage devices. Similar devices have been on the market for several years in the form of laser videodisks and audio compact disks (CDs) for consumer use. These laser videodisks are analog, that is, the disk contains one spiral track, like the track on a phonograph record. Optical disks for computer applications are digital and store their information on concentric tracks, like their magnetic cousins. Currently, three versions of optical disk technology are competing for the mass-storage market, they are read-only optical disks, write-once optical disks, and erasable optical disks.

Unlike conventional magnetic disks, read-only optical disks cannot be written on and so have the functional equivalence of read-only memory. The most popular version of read-only optical disks employs the same technology as the CD that has become popular for audio recording. The technology is digital and based on a 4¾ inch optical disk that can store 540 MB on a single side. The devices are called Compact Disk Read-Only Memories (CD-ROMs).

Write-once optical disks (also called write-once, read-mostly, or WORM) are blank disks that are recorded on by the user. To write data, a powerful beam of laser light burns tiny spots or pits into the coating that covers the surface of these disks.[4] Once burnt in, the spots are not erasable. To retrieve the data, a less powerful laser is used to read the pattern of spots and convert the patterns into audiovisual signals that can be played back on a television set. Write-once optical disks are being used to replace microfilm storage. Because optical disks have the ability to store images as well as sound, their use is quite versatile. Anything that can be digitized, such as documents, pictures, photographs, line drawings, and music, can be recorded and stored on an optical disk.

Erasable optical disks use lasers to read and write information to and from the disk but also use a magnetic material on the surface of the disk and a magnetic write head to achieve erasability. To write on such as disk, a laser beam heats a tiny spot on it; then a magnetic field is applied to reverse the magnetic polarity of the spot. Erasable optical disk systems offer the same storage capabilities of the non-erasable optical disks, along with the same reusability capabilities of conventional magnetic disks, such as Winchester systems.

Key Words

bandwidth	带宽
compact	紧密的，结实的，紧凑的
concentric	集中的，同心的
conversation	对话，会话
cursor	光标
decrement	减量，递减
differentiate	区分，区别，使有差异
equivalence	等价，等值，等效
erasability	可擦除性
horizontal	水平的，横的
increment	增加，增量，递增
interlace	隔行，交错
mouse	鼠标
non-interlace	非隔行
optical	光学的
periodically	定期地
phonograph	唱机，留声机

punctuation	标点，标点符号
responsive	响应，应答的
reusability	可重复使用的
solution	分辨率
spacebar	空格键
typewriter	打字机
versatile	多功能的，易变的
video	视频
vivid	生动的，活泼的

Notes

[1] In interlace style, electron-beam sweeps elements in odd rows first time and does elements in even rows second time.

本句中的 electron-beam 译为电子束，elements 译为像素。

译文：在隔行方式中，电子束首先扫描奇数行中的像素，第二次再扫描偶数行中的像素。

[2] By monitoring the status of the button, the system can also differentiate between clicking the button and holding it down.

本句中的"By monitoring the status of the button"是分词短语作为状语。

译文：依靠对按键状态的监测，系统也就能区分单击按键与按着按键拖曳之间的差别。

[3] Similarly, software interpretation of the mouse position means that the cursor doesn't jump completely off the screen when the mouse is moved a long distance in one direction.

本句中，"when"引导时间状语从句。

译文：同样，用软件描述鼠标位置，也意味着当鼠标沿一个方向长距离移动时，光标不会完全跳离屏幕。

[4] To write data, a powerful beam of laser light burns tiny spots or pits into the coating that covers the surface of these disks.

本句中的"To write data"是目的状语，"that covers the surface of these disks"是定语从句，修饰"the coating"。

译文：为了写入数据，激光的强大光束将覆盖在光盘的表层并烧结出小斑点或凹点。

1.4.2　Text B

A bus is a shared communication link, which uses one set of wires to connect multiple subsystems. The two major advantages of the bus organization are versatility and low cost. By defining a single connection scheme, new devices can easily be added, and peripherals can even be moved between computer systems that use the same kind of bus. Furthermore, buses are cost effective, because a single set of wires is shared in multiple ways. The major disadvantage of a bus is that it creates a communication bottleneck, possibly limiting the maximum I/O throughput. When I/O must pass through a single bus, the bandwidth of that bus limits the maximum I/O throughput.

One reason bus design is so difficult is that the maximum bus speed is largely limited by physical factors: the length of the bus and the number of devices. These physical limits prevent us from running the bus arbitrarily fast. Within these limits, there are a variety of techniques we can use to increase the performance of the bus; however, these techniques may adversely affect other performance metrics. For example, to obtain fast response time for I/O operations, we must minimize the bus latency by streamlining the communication path. On the other hand, to sustain high I/O data rates, we must maximize the bus bandwidth. The bus bandwidth can be increased by using more buffering and by communicating larger blocks of data, both of which increase the bus latency! Clearly, these two goals, low latency and high bandwidth can lead to conflicting design requirements. Finally, the need to support a range of devices with widely varying latencies and data transfer rates also makes bus design challenging.

A bus generally contains a set of control lines and a set of data lines. The control lines are used to signal requests and acknowledgments, and to indicate what type of information is on the data lines. The data lines of the bus carry information between the source and the destination. This information may consist of data, complex commands, or addresses. For example, if a disk wants to write some data into memory from a disk sector, the data lines will be used to indicate the address in memory in which to place the data as well as to carry the actual data from the disk. The control lines will be used to indicate what type of information is contained on the data lines of the bus at each point in the transfer. Some buses have two sets of signal lines to separately communicate both data and address in a single bus transmission. In either case, the control lines are used to indicate what the bus contains and to implement the bus protocol.

Each I/O device is connected to the computer system's address, data, and control buses. Each I/O device includes I/O interface circuitry. It is actually this circuitry that interacts with the buses. The circuitry also interacts with the actual I/O device to transfer data.

As for the generic interface circuitry for an input device, the data from the input device goes to the tri-state buffers. When the values on the address and control buses are correct, the buffers are enabled and data passes on to the data bus. The CPU can then read in this data. When the conditions are not right, the logic block does not enable the buffers. They are tri-stated and do not place data onto the bus. The key to this design is the enable logic. Just as every memory location has a unique address, each I/O device also has a unique address. The enable logic must not enable the buffers unless it receives the correct address from the address bus. It must also get the correct control signals from the control bus.

The design of the interface circuitry for an output device is somewhat different from that for the input device. Tri-state buffers are replaced by the register. The tri-state buffers are used in input device interfaces to make sure that no more than one device writes data to the bus at any time. Since the output devices read data from the bus, rather than write data to it, they do not need the buffers. The data can be made available to all output devices. Only the device with the correct address will read it in.

Key Words

acknowledgement	承认，确认
adversely	不利地，相反地
bottleneck	瓶颈
destination	目的，目标，目的地
latency	潜在，潜伏
scheme	计划，方案，体系，大纲
separately	分别地，个别地
streamline	流线，流线型
subsystem	子系统
sustain	承受，支持
transmission	传送，传达，传播
versatility	多用途，多功能，易变

1.4.3 Exercises

1. Translate the following phrases into English

(1)总线带宽

(2)总线协议

(3)电子枪

(4)像素距离

(5)输出设备

2. Translate the following phrases into Chinese

(1)communication bottleneck

(2)numeric keypad

(3)electron-beam

(4)scan style

(5)data transfer rate

3. Identify the following to be True or False according to the text

(1)The mouse can use LED to generate pulses.

(2)Keyboard, mouse and monitor are all input devices.

(3)A monitor is the best window for conversation between users and a computer.

(4)When we design a bus, we do not need to consider the bus speed.

(5)A bus generally contains a set of control lines and a set of data lines.

4. Reading Comprehension

(1)Which is wrong in below four items? _____

a. The keyboard has letter keys and punctuation keys.
b. The keyboard has spacebar and punctuation keys.
c. The keyboard has function keys and numeric keys.
d. The keyboard has not arrow keys.

(2)_____ is the distance between two picture elements in horizontal direction.
a. Element distance
b. Scan style
c. Solution
d. Vertical scan rate

(3)The _____ is used to implement the bus protocol.
a. data bus
b. control bus
c. address bus
d. signal bus

(4)Video bandwidth is related to the highest work frequency of the monitor. It is from_____.
a. five MHz to hundreds MHz
b. tens MHz to twenty MHz
c. tens MHz to hundreds MHz
d. tens MHz to thousands MHz

1.5 计算机专业英语的词汇特点

1.5.1 计算机专业英语的特点

计算机专业英语以表达计算机专业知识和技术的概念、理论和事实为主要目的，专业英语的主要特点是它具有很强的专业性。因此，在表达中要注重客观事实和真相，要求逻辑性强、条理规范、精练及正式。

计算机行业是一个充满活力和不断创新的行业。软、硬件技术方面的新思想和新概念层出不穷，最新的计算机专业软件及图书通常是用英语书写和描述的，正确理解和翻译这些新的专业知识和术语是非常重要的。

计算机专业是融科学性与技术性于一体的学科，在专业文献的表达中应当遵循科技文体的规范。因为专业科技文献所涉及的内容（如科学定义、定理、方程式或公式、图表等）一般并没有特定的时间关系，所以在专业文献中大部分都使用一般现在时。一般过去时、一般完成时也在专业英语中经常出现，如科技报告、科技新闻、科技史料等。用尽可能少的单词来清晰地表达原意，这就导致了非限定动词、名词化单词或词组及其他简化形式的广泛使用。

计算机专业英语文献中专业术语多，而且派生和新出现的专业术语还在不断地增加。

另外，计算机专业英语文献中的缩略词多，而且新的缩略词还在不断地增加，并成为构成新词的词源。为了准确、精细地描述事物过程，有时句子很长，会出现一段就是一个句子的情况。长句反映了客观事物中复杂的关系，它与前述精练的要求并不矛盾，句子虽长，但结构仍是精练的，只是包含的信息量大、准确性较高。

概括起来，计算机专业英语在用词和语法上有如下显著特点：
- 专业术语多。
- 半技术词汇多。
- 名词性词组多。
- 合成新词多。
- 介词短语多。
- 非限定动词（尤其是分词）使用频率高。
- 缩略词使用频繁。
- 常使用动词或名词演化成的形容词。
- 希腊词根和拉丁词根比例大。
- 被动语态使用频繁，而且多为没有行为的被动语态。
- 长句和多重复合句较多，句子中又嵌入句子。
- 常用 It…句型结构。
- 插图、插画、表格、公式、数字所占比例大。
- 用虚拟语气表达假设或建议。
- 经常使用名词作为定语，以取得简洁的效果。
- 在说明书、手册中广泛使用祈使语句。

以下句子能充分体现计算机专业英语的用词和语法特点。

例：This approach mitigates complexity separating the concerns of the front end, which typically revolve around language semantics, error checking, and the like, from the concerns of the back end, which concentrates on producting output that is both efficient and correct.

译文：这种方法降低了把处理语义、检测错误等前端工作和主要产生正确有效输出的后端工作分离开来的复杂性。

例：Alternative calculation models in neural networks include models with loops, where some kind of time delay process must be used, and "winner takes all" models, where the neuron with the highest value from the calculation fires and takes a value 1, and all other neurons take the value 0.

译文：神经网络的另外计算模型包括带回路模型和"胜者通吃"模型，其中带回路模型必须使用某种时间延迟处理，而在"胜者通吃"模型中，从计算中得到最大值的神经元触发并赋值 1，所有其他神经元赋值 0。

上面这两句话描述的是计算机专业领域内的知识，如果不了解编译原理、神经网络技术及其相关词汇，则很难给出准确的翻译。

例：The procedure by which a computer is told how to work is called programming.

译文：告诉计算机如何工作的过程称为程序设计。

该句的主要结构为 The procedure is called programming。采用一般现在时和被动语态。by which 为"介词+关系代词"引导定语从句，从句的谓语也为被动语态，which 指代 procedure。

例：Were there no plants, there would be no photosynthetic and life could not go on.

译文：如果没有植物，则没有光合作用，生命就无法继续下去。

在说明事理并涉及各种前提和条件时，可以用虚拟语气。

例：Backing up your files safeguards them against loss if your hard disk fails or you accidentally overwrite or delete data.

译文：当硬盘发生故障或用户意外覆盖、删除数据时，备份可以保护文件以避免损失。

例：Insert new diskette for drive A: and strike any key when ready.

译文：将新盘插入 A 驱动器，准备好后按任一键。

例：Must be structure field name.

译文：需要的是结构字段名。

祈使语句常用来表示指示、建议、劝告和命令等意思，可以用于说明书、操作规程和注意事项等资料中。

例：Written language uses a small number of symbols which are easily encoded in digital form and can be combined in innumerable ways to convey meaning.

译文：书面语言只使用少数符号，它们很容易以数字形式编码，并且可以用数不清的方法进行组合以便表达意义。

句中 are encoded 和 can be combined 是并列谓语，用被动语态，in digital form 和 in innumerable ways 中的介词 in 表示以什么形式，用什么方式。

例：This instrument works on the principle that each individual substance emits a characteristic spectrum of light when its molecules are caused to vibrate by the application of heat, electricity, etc.; and after studying the spectrum which he had obtained on this occasion, Hildebrand reported the gas to be nitrogen.

译文：这个仪器工作的原理是，当物质的分子由于加热、通电等而引起振动时，每种物质产生一种独特的光谱；在研究了此种情况下收集到的光谱后，Hildebrand 宣布这种气体是氮。

这个句子的基本骨架是用 and 连接的两个并列的主句 This instrument works on the principle 和 after studying the spectrum，Hildebrand reported the gas to be nitrogen。第一个主句有一个用 that 连接的同位语从句，说明主句中的 principle，而这个同位语从句又有一个用 when 连接的时间状语从句。第二个主句包含了一个用 which 连接的定语从句，说明 spectrum。

例：Having developed the capacity to store vast quantities of data, and compress it into a small space, and now having the capability to accurately and quickly retrieve information via adaptive pattern recognition make up the core which is taking multimedia solutions from a fringe technology to something which is at the beginning of pervading every aspect of our lives, and finally, living up to the promises that the hype has generated during the last five years.

译文：已经能够存储大量的数据并把它压缩到一个小的空间，能够迅速而准确地通过自适应模式识别来恢复信息，这些构成了这个核心。该核心正在使多媒体从一项边缘技术变成一项渗透我们生活的每个方面，并且最后将实现广告在近 5 年里所做出许诺的技术。

这是长句，翻译时要注意准确和精练。本句中的主语是动名词短语 Having developed the capacity to store vast quantities of data, and compress it into a small space, and now having the capability to accurately and quickly retrieve information via adaptive pattern recognition；谓语是 make up；宾语是 the core；which is at the beginning of pervading every aspect of our lives, and finally, living up to the promises 是定语从句，修饰 something。that the hype has generated during the last five years 也是定语从句，修饰 promises。

例：Technology for development will allow construction of larger projects, artificial intelligence (intelligent agents, knowledge based systems, data mining and intelligent filtering, and so on) will be increasingly feasible as costs decrease, performance improves and widespread networking are available.

译文：技术发展造成的成本的降低、性能的提高和网络的广泛应用使得建造更大的工程、人工智能（智能代理、知识库系统、数据挖掘及智能过滤等）将愈加可行。

这也是一个长句。本句的主语是 Technology for development，谓语是 will allow，其后是一宾语从句，作为 allow 的宾语。在该宾语从句中，construction of larger projects, artificial intelligence (intelligent agents, knowledge based systems, data mining and intelligent filtering, and so on)作主语，will be increasingly feasible 是系表结构作为谓语，as 引导的原因状语从句作为状语，修饰谓语。

例：It was in the 1940s that the first computer was built.

译文：第一台数字计算机建成于 20 世纪 40 年代。

例：It is necessary to learn Visual Basic.

译文：学习 Visual Basic 是很有必要的。

以上两个例句是 It…结构在专业英语中的应用。

1.5.2 专业英语词汇的构成特点

词汇是阅读、翻译和写作的基础。随着社会发展和信息传递的发展，新的词汇层出不穷，尤其是在计算机科学与技术领域中，这种现象更为突出。专业英语的构词有如下两个显著特点：大部分专业词汇来自希腊语和拉丁语；前缀和后缀的出现频率非常高。希腊语和拉丁语是现代专业英语词汇的基础。各行各业都有一些自己领域的专业词汇，有的是随着本专业发展应运而生的，有的是借用公共英语中的词汇，有的是借用外来语言词汇，有的则是人为构造成的词汇。

计算机专业英语中常见的词汇类型有以下几种。

1. 技术词汇和次技术词汇

技术词汇的意义狭窄、单一，一般只使用在各自的专业范围内，因而专业性很强。这类词一般较长，并且越长，词义越狭窄，出现的频率也不高。

例如：

superconductivity（超导性），hexadecimal（十六进制），amplifier（放大器），bandwidth（带宽），flip-flop（触发器）等。

次技术词汇是指不受上下文限制的各专业中出现频率都很高的词。这类词汇往往在不同的专业中具有不同的含义。

例如：register 在计算机系统中表示寄存器，在电学中表示计数器、记录器，在音乐中表示音区，而在日常生活中则表示登记簿、名册、挂号信等。

2．合成词

合成词是专业英语中另一大类词汇，其组成面广，多数以连接符"-"连接单词构成，或者采用短语构成。合成方法有名词+名词、形容词+名词、动词+副词、名词+动词、介词+名词、形容词+动词等。但是，合成词并非随意可以构造，否则会形成一种非正常的英语句子结构，虽然可由多个单词构成合成词，但这种合成方式太冗长，应尽量避免。

下面是以"-"连接的合成词。

file + based → file-based　基于文件的
Windows + based → Windows-based　以 Windows 为基础的
object + oriented → object-oriented　面向对象的
thread + oriented → thread-oriented　面向线程的
point + to + point → point-to-point　点到点
plug + and + play → plug-and-play　即插即用
pear + to + pear → pear-to-pear　对等的

随着词汇的专用化，合成词中间的连接符被省略，形成了一个独立的单词。例如：

in + put → input　输入
out + put → output　输出
feed + back → feedback　反馈
on + line → online　在线

在英语中，有很多专业术语由两个或更多的词组成，称为复合术语。它们的构成成分虽然看起来是独立的，但实际上合起来构成一个完整的概念。

liquid crystal　　　　　　　液晶
computer language　　　　　计算机语言
civil engineering　　　　　　土木工程

3．派生词

派生词法也称为缀合，这类词汇非常多，专业英语词汇大部分都是用派生法构成的。它是根据已有的词汇，通过对词根加上各种前缀和后缀来构成新词。这些词缀有名词词缀，如 inter-、sub-、in-、tele-、micro-等；形容词词缀，如 im-、un-、-able、-al、-ing、-ed 等；动词词缀，如 re-、under-、de-、-en、con-等。其中，采用前缀构成的单词在计算机专业英语中占了很大比例。

英语的前缀是有固定意义的，记住其中的一些常用前缀对于记忆生词和猜测词义有很大帮助。加前缀构成新词只改变词义，不改变词性。加后缀构成新词可能改变也可能不改变词义，但一般改变词性。有的派生词加后缀时，语音或拼写可能发生变化。从一个词汇的后缀可以判别它的词类，这是它的语法意义。它们的词汇意义往往并不明显。

下面是一些典型的派生词。

（1）hyper-　超级
hypertext　超文本　　　　　　　　　　hypermedia　超媒体
hyperswitch　超级交换机　　　　　　　hypersonic　超音速的
（2）inter-　相互，在……之间
interface　接口　　　　　　　　　　　intercommunication　相互通信
interlace　隔行扫描　　　　　　　　　interactive　交互的
（3）micro-　微型
microprocessor　微处理器　　　　　　 microelectronics　微电子
microcomputer　微型计算机　　　　　　microwave　微波
（4）multi-　多，多的
multimedia　多媒体　　　　　　　　　 multiprocessor　多处理器
multiprogram　多道程序　　　　　　　 multicast　多点传送
（5）poly-　多，复，聚
polycrystal　多晶体　　　　　　　　　polytechnical　多工艺的
polyatomic　多原子的　　　　　　　　 polymorphism　多态性
（6）re-　再，重新
rerun　重新运行　　　　　　　　　　　rewrite　改写
resetup　重新设置　　　　　　　　　　reexchange　再交换
（7）semi-　半
Semiconductor　半导体　　　　　　　　semiautomatic　半自动的
semidiameter　半径　　　　　　　　　 semicircular　半圆的
（8）super-　超级
supercomputer　超级计算机　　　　　　superclass　超类
superstructure　上层建筑　　　　　　 superuser　超级用户
（9）tele-　远程的，电的
telephone　电话　　　　　　　　　　　teleconference　远程会议
telescope　望远镜　　　　　　　　　　telegraph　电报
（10）ultra-　超过，极端
ultrashort　超短（波）的　　　　　　 ultrared　红外线的
ultraspeed　超高速的　　　　　　　　 ultramicroscope　超显微镜
（11）un-　反，不，非
unformat　未格式化的　　　　　　　　 undelete　恢复
uninstall　卸载　　　　　　　　　　　unimportant　不重要的
（12）-able　可能的
programmable　可编程的　　　　　　　 portable　便携的
adjustable　可调整的　　　　　　　　 considerable　值得重视的

（13）-ate 成为……，用……处理
eliminate 消除
terminate 终止
circulate 循环，流通
estimate 估计，估算
（14）-ic 有……特性的，属于……的
academic 学术的
atomic 原子的
elastic 灵活的
periodic 周期的
（15）-ive 有……性质的，与……有关的
productive 生产的
active 主动的
expensive 昂贵的
attractive 有吸引力的
（16）-ize ……化，变成
characterize 表示……的特性
optimize 完善
industrialize 使工业化
realize 实现
（17）-lity ……性能
reliability 可靠性
confidentiality 保密性
（18）-ment 行为，状态
development 发展
equipment 设备
agreement 同意，协议
adjustment 调整
（19）-meter 计量仪器
barometer 气压表
telemeter 测距仪
（20）-ware 件，部件
hardware 硬件
software 软件

4. 借用词

借用词是指借用公共英语及日常生活用语中的词汇来表达专业含义。借用词一般来自厂商名、商标名、产品代号名、发明者名、地名等，也可将普通公共英语词汇演变成专业词义而实现。也有对原来词汇赋予新的含义的。例如：

cache 高速缓存
register 寄存器
mailbomb 邮件炸弹
firewall 防火墙
flag 标志，状态
semaphore 信号量

英语科技文体中有很多词汇并不是专业术语，但在日常口语中用得不是很多，它们多见于书面语中。掌握这类词对阅读科技文献或写科技论文十分重要。例如：

accordance 按照
alternative 交替的
appropriate 恰当的
compensation 补偿
confirm 证实
inclusion 包括
induce 归纳
nonetheless 然而
acknowledge 承认
application 应用
circumstance 情况
imply 隐含
modification 修改
indicate 指示
initial 初始的
nevertheless 然而

5．通过词类转化构成新词

这是指一个词不变化词形，而由一种词类转化为另一种或几种词类，有时发生重音或尾音的变化。英语中名词、形容词、副词、介词可以转化成动词，而动词、形容词、副词、介词可以转化成名词。但是，最活跃的是名词转化成动词和动词转化成名词。

例如：
but（连词）但是→ but（介词）除了
by（介词）在……旁边→ by（副词）在一旁
center（名词）中心→ center（动词）集中
clear（形容词）明确的→ clear（动词）清除
close（动词）关上→ close（副词）靠近
coordinate（动词）协调→ coordinate（名词）坐标
hard（形容词）坚硬的→ hard（副词）努力地

名词化是英语科技文体中一个常见的现象。所谓名词化是指把动词变成有动作含义的名词。如果是动词短语或者是句子，则把这个动词短语或句子变成名词短语。

"air moves"可以转换成"the motion of air"，其含义为"空气运动"，air 是 motion 的行为主体。

"to apply force"可以转换成"the application of force"，其含义为"应用力"，force 是 application 的宾语。

"analytical chemists develop the equipment"可以转换成"the development of the equipment by analytical chemists"，其含义为"由分析化学家开发的设备"，analytical chemists 是 the development 的行为主体。

6．词汇缩略

词汇缩略是指将较长的单词取其首部或主干构成与原词同义的短单词，或者将组成词汇短语的各单词的首字母拼接为一个大写字母的字符串。对于计算机专业来说，在专业文献、程序语句、软件文档、互联网信息和文件描述中大量采用词汇缩略。

词汇缩略有以下三种形式。

◆ 节略词

某些词汇在发展过程中为方便起见逐渐用它们的前几个字母来表示，这就是节略词；或者在一个词组中取各词的一部分，重新组合成一个新词，表达的意思与原词组相同。例如：

maths——mathematics　　　　　　　　　数学
ad——advertisement　　　　　　　　　　广告
kilo——kilogram　　　　　　　　　　　公斤
dir——directory　　　　　　　　　　　目录
lab——laboratory　　　　　　　　　　　实验室
radar——radio detection and ranging　　雷达
transceiver——transmitter receiver　　　收发信机
TELESAT——Telecommunications satellite　通信卫星

◆ 缩写词

缩写词并不一定由某个词组的首字母组成,有些缩写词仅由一个单词变化而来,而且大多数缩写词的每个字母或最后一个字母后都附有一个句点。例如:

e.g. —— for example
Ltd. —— limited
sq. —— square

◆ 缩略词和首字词

缩略词是指由某些词组的首字母所组成的新词。例如:

ROM——Read Only Memory(只读存储器)
RAM——Random Access Memory(随机存取存储器)
COBOL——Common Business Oriented Language(面向商务的通用语言)

首字词与缩略词基本相同,区别在于首字词必须逐字母念出。例如:

CPU——Central Process Unit(中央处理器)
BBS——Bulletin Board System(电子公告板系统)
FTP——File Transfer Protocol(文件传输协议)
DBMS——DataBase Management System(数据库管理系统)
HTTP——Hypertext Transfer Protocol(超文本传输协议)
SMTP——Simple Mail Transfer Protocol(简单邮件传输协议)
CGA——Color Graphics Adapter(彩色图形适配器)

1.5.3 专业术语与命令

在计算机语言、程序语句、程序文本注释、系统调用、命令字、保留字、指令字及网络操作中广泛使用专业术语进行信息描述。随着计算机技术的发展,这样的专业术语还会进一步增加。

1. Internet 地址

域名(Domain Name)是 Internet 中主机地址的一种表示方式。域名采用层次结构,每层构成一个子域名,子域名之间用点号隔开且从右到左逐渐具体化。域名的一般表示形式为:计算机名、网络名、机构名、一级域名。对一级域名有一些规定,用于区分机构和组织的性质,例如 edu 表示教育机构、com 表示商业单位、mil 表示军事部门、gov 表示政府机关、org 表示其他组织。

用于区分地域的一级域名采用标准化的 2 个字母的代码。例如:

cn	中国	ca	加拿大
us	美国	au	澳大利亚
gb	英国(官方)	uk	英国(通用)
tw	中国台湾	hk	中国香港
fr	法国	un	联合国
nz	新西兰	dk	丹麦

在 Internet 上,电子邮件(E-mail)地址具有如下统一的标准格式:用户名@主机域名。

例如，zhang_1@online.sh.cn 是一个电子邮件的地址，其中的 zhang_1 是用户名，@是连接符，online.sh.cn 是"上海热线"的主机域名，这是注册"上海热线"后得到的一个 E-mail 地址。

2. 专用的软件名称

人类相互交流信息所用的语言称为自然语言，但当前的计算机还不能理解自然语言，它能理解的是计算机语言，即软件。软件分成系统软件和用户软件。近年来，随着计算机技术的发展，新的软件不断推出。下面是一些常用软件的名称。

NextStep　面向对象操作系统（Next）
Netware　局域网络操作系统（Novell）
Daytona　视窗型操作系统（MS）
Java　网络编程语言（Sun）
Excel　电子表格软件（MS）
Informix　关系数据库系统（Informix）
Navigator　互联网浏览软件（Netscape）

3. 专用计算机厂商及商标名

下面给出的是一些著名计算机公司的译名。

Microsoft　微软　　　　　Philip　飞利浦
HP　惠普　　　　　　　　DELL　戴尔
Panasonic　松下　　　　　Hisense　海信
Acer　宏碁　　　　　　　Intel　英特尔

4. 命令和指令

每个处理器都具有很多指令，每台机器也具有很多系统命令，不同的操作系统也定义了不同的操作命令，它们通常是缩写的。牢记这些指令，就熟悉了计算机的操作；了解缩写的含义，也就了解了所用的操作的含义。例如：

创建目录　MD（make directory）
改变目录　CD（change directory）
删除目录　RD（remove directory）
列表目录　DIR（directory）
重命名　REN（rename）
取偏移地址指令　LEA（load effective address offset）
取数据段地址指令　LDS（load doubleword pointer）
总线封锁命令　LOCK（assert bus lock signal）
串装入操作指令　LODS（load string operand）
中断请求　INT（call to interrupt procedure）
中断返回　IRET（interrupt return）

5. 屏幕信息

目前，大部分软件都采用了菜单技术及其他人机对话技术。能否正确地阅读和理解这些屏幕信息，关系到我们能否正确使用这些软件，以及充分发挥软件所提供的全部功能。很多软件都附有帮助信息的相关文件供用户调用，从而帮助用户掌握软件的功能。常见的文件有 Readme、Message、Help、Assist 等。

当一个程序具有若干供用户选择的功能时，通常使用交互技术进行分支处理。实现的过程是：屏幕首先显示出提供的功能名称，用户根据需要指出希望完成的功能，然后由程序分析用户的选择并调用不同的功能块进行处理。此过程称为"菜单技术"（Menu technique）。

为了使程序具有较好的通用性、坚固性及人机交互性，在设计屏幕"菜单"和其他输入/输出内容时，应符合下述主要原则：
- 输入方式要简单，并尽可能每步给出屏幕提示。
- 屏幕显示信息应简洁、易懂，避免二义性。
- 尽可能在一屏中包含更多的信息。
- 检测所有输入数据的合法性。
- 尽量减少用户处理出错的工作量。
- 如果用户选择的功能可能会产生严重的后果（如删除文件、格式化磁盘等），应再次予以确认，以提醒用户不致误操作。

1.6 习题答案与参考译文

1.6.1 第 1 单元习题答案

1.1 Hardware Basics

1. Translate the following phrases into English

(1) 逻辑电路　　　　　　　　logic circuit
(2) 输出变量　　　　　　　　output variable
(3) 二进制信息　　　　　　　binary information
(4) 组合电路　　　　　　　　combinational circuit
(5) 正向脉冲　　　　　　　　positive pulse

2. Translate the following phrases into Chinese

(1) Boolean Algebra　　　　　布尔代数
(2) chief executive officer　　首席执行官
(3) full-time work　　　　　　全职工作
(4) Harvard College　　　　　哈佛大学
(5) scientific research program　科学研究项目

3. Identify the following to be True or False according to the text

T　F　F　T　F

4. Reading Comprehension

(1) c　temperature

(2) a   algorithm
(3) b   binary
(4) d   truth table

1.2 Central Processing Unit

1. Translate the following phrases into English

(1)取指—译码—执行	fetch-decode-execute
(2)算术/逻辑运算	Arithmetic/Logical Operations
(3)微电子技术	microelectronic techniques
(4)印制电路板	printed circuit board
(5)外围设备	peripheral equipment

2. Translate the following phrases into Chinese

(1)current instruction	当前指令
(2)instruction register	指令寄存器
(3)program counter	程序计数器
(4)retrieve information	检索信息
(5)high-speed storage	高速存储器

3. Identify the following to be True or False according to the text

T   T   F   T   F

4. Reading Comprehension

(1) c   a control unit and an arithmetic/logic unit
(2) d   instructions
(3) a   A register
(4) b   fetch-decode-execute

1.3 Memory

1. Translate the following phrases into English

(1)易失性存储器	volatile memory
(2)实模式	the real mode
(3)寻址能力	address ability
(4)闪存	flash memory
(5)刷新电路	refresh circuitry

2. Translate the following phrases into Chinese

(1)software interrupt	软件中断
(2)expanded memory	扩展存储器
(3)refresh cycle	刷新周期
(4)Read Only Memory	只读存储器

(5) Random Access Memory　　　随机存取存储器

3. Identify the following to be True or False according to the text

T　F　T　T　F

4. Reading Comprehension

(1) a　1,048,576 bytes

(2) c　2^n different bit combinations

(3) b　lost

(4) c　DRAM

1.4　Input/Output Systems

1. Translate the following phrases into English

(1) 总线带宽　　　　　　　bus bandwidth

(2) 总线协议　　　　　　　bus protocol

(3) 电子枪　　　　　　　　electron gun

(4) 像素距离　　　　　　　element distance

(5) 输出设备　　　　　　　output device

2. Translate the following phrases into Chinese

(1) communication bottleneck　　通信瓶颈

(2) numeric keypad　　　　　　 数字键盘

(3) electron-beam　　　　　　　电子束

(4) scan style　　　　　　　　　扫描类型

(5) data transfer rate　　　　　　数据传输速率

3. Identify the following to be True or False according to the text

T　F　T　F　T

4. Reading Comprehension

(1) d　The keyboard has not arrow keys.

(2) a　Element distance

(3) b　control bus

(4) c　tens MHz to hundreds MHz

1.6.2　第1单元参考译文

1.1　硬　件　基　础

1.1.1　课文 A
组合电路

　　组合电路是具有一组输入和输出信号的若干逻辑门的连接排列。在任何给定的时间，输出的二进制值是输入二进制值组合的函数。组合电路的框图如图 1-1 所示。n 位二进制输入变量来源于外部，m 位二进制输出变量也输出到外部部件，在两者之间是一个逻辑门的互连网络。组合电路通过传输二进制信息，使得给定的输入数据产生了所需的输出数据。

图 1-1　组合电路的框图

组合电路可以由表示 n 位输入变量和 m 位输出变量之间二进制位关系的真值表来描述。真值表列出了针对 2^n 个输入组合中的每种情况的相应输出二进制位的组合。组合电路也能规定 m 种布尔函数，每种函数对应一个输出变量。每个输出函数根据 n 位输入变量来表达。

逻辑系统

在直流逻辑或电平逻辑系统中，比特用两种电压电平中的一种来实现。如果较高的电压为 1 电平，而另一个是 0 电平，则称该系统使用的是直流正逻辑。另一方面，把比特的较低的电压状态记为 1 电平，较高的电压状态记为 0 电平，这样的系统称为直流负逻辑系统。应该说明的是在这些定义中，两个电压的绝对值是没有意义的。尤其是 0 状态并不一定表示 0 电压电平（虽然在某些系统中可能是这样）。

实际器件的各个参数对于不同样品来说是不一样的，并且这些参数随温度变化而变化。此外，电源和地线中还可能有电压脉动和电压尖峰，电路中还可能有其他不需要的被称为噪声的信号源。由于这些原因，我们不能把数字电平规定得太死，而是规定每个状态都在指定电平附近的一段电压范围内，例如 4V±1V 及 0.2V±0.2V。

在动态逻辑或脉冲逻辑系统中，通过脉冲的有无来识别比特。在动态正逻辑系统中，1 表示存在正脉冲；而在负逻辑系统中，负脉冲表示为 1。不论是哪一种系统，某一时刻在某一输入端（或输出端）的 0，总表示此刻无脉冲。

逻辑电路

数字计算机的设计基于称为布尔代数的逻辑方法学，它采用三种基本运算：逻辑加，称为"或"功能；逻辑乘，称为"与"功能；逻辑求补，称为"非"功能。布尔代数中的变量是二进制值，也就是说，一个操作或一系列操作之后的结果变量也只能是 1 或 0 这两个值之一。这两个值也可被认为是正确或错误、是或否，以及正的或负的。

因为开关只能是"关"或"开"两个状态，所以它最适合表示两状态的变量值。

只有三种基本逻辑运算：合取（逻辑乘积）通常称为"与"；析取（逻辑和）通常称为"或"；而否定通常称为"非"。

1.1.2　课文 B

比尔·盖茨是美国商业巨人、计算机程序设计员及慈善家。盖茨是世界上最大的个人计算机软件公司即微软公司的前首席执行官和现任主席。他与保罗·艾伦一起创建了微软公司。他一直名列世界富人榜，并且从1995年到2007年连续蝉联世界首富。在微软公司的职业生涯中，盖茨担任过首席执行官和首席软件架构师，是该公司最大的个人股东，持有该公司8%的普通股。他还撰写和与其他人合著了若干本书。

盖茨是个人计算机改革的著名企业家之一。在他职业生涯中的最后阶段，盖茨不遗余力地致力于慈善事业，通过"比尔及梅琳达·盖茨基金会"为各种慈善组织和科学研究项目投入大量资金，这个基金成立于2000年。

比尔·盖茨出生于华盛顿州的西雅图。他的父亲是一位著名律师，其母亲是第一洲际银行系统的董事。13岁时，他就读于独立预备学校湖滨中学。盖茨于1973年从湖滨中学毕业，他在SAT（美国大学入学考试）测试中得到1590分，满分为1600分。他于1973年秋季进入哈佛大学学习。在哈佛大学时，盖茨遇到了继他之后成为微软公司首席执行官的史蒂夫·鲍尔默。

在大学二年级的时候，盖茨为煎饼排序开发了一个求解问题的算法，这是在综合课上由他的教授哈里·刘易斯提出的一系列未解问题之一。盖茨的答案破了长达30年的快速求解纪录。后来的纪录保持者只快了1%。他的求解后来发表在与哈佛计算机科学家Christos Papadimitriou联合出版的论文上。

煎饼排序是一种变异的排序问题，只有被允许的操作才能翻转序列中前面部分的元素。它不像传统的排序算法，试图尽可能地采用最少的比较次数来排序，目的是以尽量少的交换排好序列。这种操作可被形象化地理解为一个煎饼栈，在该栈中允许将上面的k个煎饼取出并翻转它们。一个引申的问题是有关烤煎饼的，每个煎饼有一个煎好的面，此外，在结束时所有煎饼必须将煎好的面朝上。

作为哈佛的学生，盖茨并没有明确的学习计划，而是花费了大量时间使用学校的计算机。盖茨和保罗·艾伦一直保持联系，并于1974年的夏天和他一起进入了Honeywell。次年基于英特尔8080中央处理器的MITS Altair 8800发布了，盖茨和艾伦把这看成开创他们自己的计算机软件公司的机遇。这一次，盖茨从哈佛退学了。他和父母谈起过这个决定，他们在知道盖茨多么想成立这个公司后支持了他的决定。

在阅读了于1975年1月出版的介绍Altair 8800的《大众电子》杂志后，盖茨与新型微型计算机的开发者Micro Instrumentation and Telemetry Systems（MITS）公司取得了联系，告诉他们说他和伙伴们正为该平台开发BASIC语言解释程序。实际上，盖茨和艾伦既没有Altair，也没有为它写代码。他们只是想试探一下MITS公司的兴趣。MITS公司总裁爱德华·罗伯茨答应看看他们的演示。在接下来的几周内，他们开发了一个运行在微型计算机上的Altair模拟器及BASIC解释程序。

在位于阿尔伯克基的MITS办公室进行的演示非常成功，促使他们与MITS公司签署协议以发布Altair BASIC解释程序。在1975年的11月，保罗·艾伦受雇于MITS公司，盖茨从哈佛退学并与艾伦一起在位于阿尔伯克基的MITS公司工作。他们命名这种合伙企业为"微软"，并拥有了在阿尔伯克基的他们的第一间办公室。不到一年的时间，在1976年11月26日，商标名称"微软"在新墨西哥州事务机构注册。盖茨再也没有回到哈佛大学完成他的学业。

盖茨在2000年1月辞去微软公司首席执行官职务，但仍然担任微软公司主席和新创建的首席软件架构师职位。在2006年6月，盖茨宣布他将从微软公司的全职工作中过渡到兼职工作，而全职工作于"比尔及梅琳达·盖茨基金会"。

1.2 中央处理器

1.2.1 课文 A

计算机的基本工作是处理信息。为此，计算机可以定义为接收信息的装置，信息是以称为程序的指令序列和称为数据的字符形式出现的。该装置可对信息进行算术和逻辑运算，然后提供运算结果。程序的作用是指示计算机如何工作，而数据则为解决问题提供所需要的信息，两者都存储在存储器里。

人们认为计算机具有很多显著的功能。不过大多数计算机，无论是大型机还是小型机，都具有三个基本性能。

第一，计算机具有进行加、减、乘、除及求幂等各种算术运算的电路。

第二，计算机具有与用户通信的功能。如果我们不能输入信息和取出结果，那么这种计算机毕竟不会有多大用处。

第三，计算机具有进行判定的电路。电路能对如下事件做出判定：一个数是否小于另一个数？两个数是否相等？一个数是否大于另一个数？

CPU 可以是一个单独的微处理器芯片、一组芯片或者一个带有晶体管、芯片、导线和连接点的插件板。在 CPU 方面的差别可以区分大型、小型和微型计算机。处理器由两个称为控制单元和算术/逻辑单元的功能部件与一组称为寄存器的特殊工作区组成。

控制单元

控制单元是负责监督整个计算机系统操作的功能部件。在某些方面，它类似于智能电话交换机，因为它将计算机系统的各功能部件连接起来，并根据当前执行程序的需要控制每个部件完成操作。控制单元从存储器中取出指令，并确定其类型或对之进行译码，然后将每条指令分解成一系列简单的、很小的步骤或动作。这样，就可以控制整个计算机系统的逐步操作。

算术/逻辑单元

算术/逻辑单元（ALU）是为计算机提供逻辑及计算能力的功能部件。控制单元将数据送到算术/逻辑单元中，然后由算术/逻辑单元完成执行指令所需要的算术或逻辑运算。算术运算包括加、减、乘、除。逻辑运算完成比较，并根据结果采取行动。例如，比较两个数是否相等，如果相等，则继续处理；如果不等，则停止处理。

寄存器

寄存器是处理器内部的存储单元。控制单元中的寄存器用来跟踪运行程序的所有状态，它存储当前指令、下一条将执行指令的地址及当前指令的操作数这样一些信息。在算术/逻辑单元中，寄存器存放要进行加、减、乘、除及比较的数据项。而其他寄存器则存放算术及逻辑运算的结果。

指令

　　指令由操作码和操作数组成，操作码指明要完成的操作功能，而操作数则表示操作的数据。例如，如果一条指令要完成两数相加的操作，它就必须知道：这两个数是什么？这两个数在哪里？当这两个数存储在计算机的存储器中时，应有指明其位置的地址，如果一个操作数指向计算机存储器中的数据，则该操作数称为地址。处理器的工作就是从存储器中找到指令和操作数，并执行每个操作。完成这些工作后，就通知存储器送来下一条指令。

　　CPU 以一系列步骤执行每条指令：

1．从存储器取出一条指令，送入指令寄存器。
2．修改程序计数器以指向下一条指令。
3．确定刚取出指令的类型。
4．如果指令使用存储器中的数据，则须确定它们的地址。
5．取出数据并送到 CPU 内部寄存器。
6．执行指令。
7．将结果存储到适当的位置。

返回到第 1 步，开始执行下一条指令。

　　这个顺序执行的步骤常称为"取指—译码—执行"周期。所有计算机的操作都是以此为中心的。处理器以惊人的速度一遍又一遍地重复以上这一步步的操作。一个称为时钟的计时器准确地发出定时电信号，该信号为处理器工作提供有规律的脉冲信息。计量计算机速度的术语引自电气工程领域，称为兆赫（MHz）。兆赫意指百万个周期数每秒。

1.2.2　课文 B

　　微型计算机上的 CPU 实际上是一个很小的集成电路或芯片。虽然大多数 CPU 芯片比一个眼镜片还小，但所包含的电子元件在几十年前却要装满一个房间。应用先进的微电子技术，制造者能够把数万个电子元件集成到很小的可靠工作的分层硅芯片上，这些芯片的能耗很低。CPU 协调计算机各个部件的所有活动。它确定应该以什么顺序执行哪些操作。CPU 也可检索存储器的信息并将操作结果存到存储器单元里，以备以后参考。

　　微型计算机或简称微机是计算机的一种。它诞生于 20 世纪 70 年代初期。计算机的大脑称为微处理器。它是计算机内的主要芯片，可完成所有工作。它也是母板上的工作中心。它解释和执行构成计算机程序的指令。CPU 由一个算术单元及其相关电路，称为算术/逻辑单元，一个指令计数器和译码器组成的。CPU 一次只能完成一个操作。实质上，数字化编码的指令存储于计算机的高速存储器或主存中，CPU 一次提取并执行一条指令。数字化编码的指令告诉 CPU 完成什么操作，以及操作所需的数据存储在什么地方。

　　称为微处理器的微机中央处理器是单片半导体部件。也就是说，实现计算机所有逻辑和算术功能所必不可少的成千上万个分立的电路元件都制造在一个芯片上。完整的微机系统由微处理器、存储器和外围设备组成。处理器、存储器和外围设备的电子控制装置通常一起放在一个或几个印制电路板上。使用微处理器的系统可以组合起来完成迄今为止只有小型计算机系统才能够做的工作。一般来说，微机的指令系统比小型机略为简单，灵活度稍低，而且速度比小型机明显慢得多。同样，小型机可以配置较大的内存容量。但是微机的功能变得越来越强，并与小型机技术结合起来了。

实质上，微处理器是一台小计算机，它可完成基本计算器所做的工作，即对存储在计算机内存的数据进行加、减、乘、除运算。计算机程序告诉微处理器做什么，这就是 PC 内的每件事的工作方式。

微处理器的其他叫法有处理器、中央处理器（CPU），以及由微处理器的编号命名 8088、80286、80386、80486 等。PC 使用的微处理器主要有三种：8088/8086、80286 或 AT 微处理器和 386 系列处理器。不会有 586 微处理器，因为生产微处理器的 Intel 公司不再用编号命名，而是称其为奔腾（Pentium）微处理器。正是微处理器的这个"微"字才使一些老计算机用户称 PC 为微机。这可能适用于第一代微处理器，而目前发电站用的 PC 根本不"微"。

1.3 存 储 器

1.3.1 课文 A

一个存储器单元是一个电路，或在某些情况下只是一个能存储 1 位信息的单个器件。存储器单元的系统排列组成了存储器。存储器也必须包括外围电路，以此来寻址，并将数据写到单元内，以及检测存储在单元中的数据。

分为两类基本的半导体存储器。第一类是随机存取存储器（RAM），是一种可读可写存储器，它的每个独立单元可以在任何指定的时间内寻址。对每个单元的访问时间实际上是一样的。RAM 的定义指出，每个单元都允许做读和写的操作，读和写所用的访问时间几乎相同。

第二类半导体存储器是只读存储器（ROM）。尽管这类存储器中所设置的数据在某些设计中可以改变，但这些数据通常是固定的。不过，在 ROM 中写一个新数据所需要的时间比对存储器单元的读取访问时间要长得多。例如，ROM 可用于存储系统操作程序的指令。

易失性存储器是一种当电路中电源断开时数据丢失的存储器，而非易失性存储器是即使断开电源也能保存数据的存储器。一般，随机存取存储器是一种易失性存储器，而只读存储器是非易失性存储器。

RAM 的两种类型是静态 RAM（SRAM）和动态 RAM（DRAM）。静态 RAM 由基本的双稳态触发器电路组成，它只需要直流电流或电压以保持记忆。静态 RAM 有两个稳定状态，定义为逻辑 1 和逻辑 0。动态 RAM 是 MOS 存储器，当在一个电容上充电时，它存储 1 位信息。由于电容上的电荷会延迟一个固定的时间常数（ms，毫秒级），所以需要定期刷新以恢复充电从而使动态 RAM 不丢失它存储的信息。

SRAM 的优点是这个电路不需要额外复杂的刷新周期和刷新线路，但它的缺点是电路相当大。一般，SRAM 的一个位需要 6 个晶体管。DRAM 的优点是它只由一个晶体管和一个电容组成，但缺点是需要刷新线路和刷新周期。

ROM 一般有两种类型。第一类既可以由制造厂家编程（掩模可编程的）也可以由用户编程（可编程的或 PROM）。一旦 ROM 采用了任何一种方法编程，存储器中的数据就固定了且不能改变。第二类可以认为是一种可改变的 ROM，如果需要的话，对 ROM 中的数据可以重新编程。这类 ROM 可以称为 EPROM（可擦除的可编程的 ROM），EEPROM（可电擦除的 PROM）或闪存。正如在前面提到的那样，对这些存储器中的数据可以重新编程，尽管所包含的时间远远长于读取访问时间。在某些情况下，在重新编程过程中，有可能不得不从电路中移走存储器芯片。

基本的存储器结构如图 1-2 所示。端点的连接可以包括输入、输出、地址、读和写控制。存储器的主要部分是数据存储体。一个 RAM 存储器将包括上面提到的所有端点的连接，而一个 ROM 存储器不包括输入和写控制。

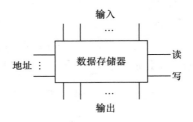

图 1-2　基本的存储器结构

计算机内存以信息的千字节或兆字节来度量。1 字节等于存储一个字符，如一个字母或数字。1KB 等于 1024 字节，1MB 约等于 1 048 576 字节。软件需要一定数量的内存来正常工作。如要给计算机增加新的软件，在软件包装上通常可以找到该软件所需要的确切内存容量。

存储器由许多单元组成，每个单元可以存储一块信息。每个存储单元有一个号码，称为单元地址。通过地址，程序可以访问存储单元。假定存储器有 n 个单元，它们就有地址编号 $0 \sim n-1$。存储器的所有单元具有同样的位数。如果一个单元有 k 位，则它可以保存 2^k 个不同位组合数据中的任一个。注意相邻的单元具有连续的地址。

使用二进制值系统的计算机，也用二进制值表示存储器地址。如果地址有 m 位，可直接寻址的最大单元数量是 2^m 个。地址的位数与存储器可直接寻址的最大单元数量有关，而与每个单元的位数无关。具有 8 位长的 2^{12} 个单元的存储器和具有 64 位长的 2^{12} 个单元的存储器都需要 12 位地址。

单元的含义表示它是最小的可寻址单位。近年来，大多数计算机制造商已经使单元的长度标准化为 8 位，这样的单元称为字节。字节可组成字，16 位字长的计算机的每个字包含 2 字节，而 32 位字长的计算机的每个字则包含 4 字节。字的含义是大多数指令对整字进行操作，比如把两个字相加在一起。因而 32 位机器有 32 位的寄存器和指令，以实现传送、加法、减法和其他 32 位字的操作。

1.3.2　课文 B

20 世纪 70 年代，计算机有了进一步的发展，使计算机领域发生了一场革命。这就是将成千上万个集成电路蚀刻在一小块硅（芯）片上的能力，硅（芯）片是具有半导体特性的非金属元件。芯片上具有成千上万个相同的电路，每个电路能存储 1 位。由于芯片很小，并且电路蚀刻在芯片上，电信号无须行进很远，因此它们传输得较快。此外，装有电路的部件体积可以大大减小，这一进步导致了小型机和微型机的引入。其结果是计算机体积变小，速度加快，价格更便宜。但半导体存储器有一个问题，当切断电源时，存储器里的信息就丢失了，而不像磁芯存储器那样在断电时还能保留信息。

在实模式下运行的 80x86 系列处理器，其物理寻址能力达到 1 兆字节内存。EMS 采用页面调度或存储切换技术，允许实模式处理访问附加的存储空间。为了扩展存储器，需要额外的硬件和驱动程序。存储切换寄存器充当有 1 兆字节空间的物理窗口和驻留在扩展存储板上的逻辑存储器之间的网关。设备驱动器也称为扩展内存管理器（EMM），控制寄存器，使得程序的存储器存取可以重定位在整个可用的扩展存储器上。

为了访问扩展存储器，程序需要与 EMM 通信。与 EMM 通信的方式与调用 DOS 类似。程序中设立了专门的 CPU 寄存器并建立了一个软件中断请求，同时定义了 30 多个功能，并用应用程序和操作系统来控制扩展存储器。当一个程序分配了扩展存储器页时，EMM 就将一个标志回复给这个请求程序。当再次调用 EMM 时，这个标志将用来区分逻辑页中哪些块正在使用。

多种不同类型的高速缓存（磁盘高速缓存、内存高速缓存、处理器高速缓存）都能改善系统总性能。尽管许多高档的系统在系统设计中就包含了高速缓存，但是，从低档 8086 系统到以 i486 为基础配置的最高性能系统，高速缓存都可以作为选件加以实现。

有两种通用的方法可实现磁盘高速缓存。这两种方法的主要区别在于用作高速缓存的存储器位于什么地方。实现磁盘高速缓存的第一种即最常用的方法是使用扩展内存。在一台微机中，1MB 或者更大的扩展内存可分配作为磁盘高速缓冲存储器。"命中"率（指对磁盘的数据读取能从高速缓存中读出，而不是从磁盘中读出）越高，系统总性能就越好。对于文字处理或管理小的文件，一个大的高速缓存可能用处不大。对于管理复杂图形的大的数据库文件，利用扩展内存作高速缓存，可明显地提高性能。实现磁盘高速缓存的第二种方法是采用高速缓存控制器。这些控制器不仅能控制联机的硬盘驱动器的读/写操作，它们还提供很多基于系统的扩展存储器的高速缓存的功能。然而，把高速缓存置于控制器上，高速缓存的性能可以比多数使用扩展内存作高速缓存的性能好得多。

在高速缓存控制器上采用了高速存储器，当系统调用硬盘时，高速缓存中的控制器确定被调用的数据是在高速缓存中，还是在磁盘上。当使用扩展内存作高速缓存时，高速缓存截获对硬盘进行读操作的调用，以检查被系统调用的数据是否在高速缓存中。如果数据不在高速缓存中，就向磁盘控制器发送一个读指令。这个处理过程要花时间。这不仅仅是用于高速缓存的扩展存储器的速度要比高速缓存控制器中的存储速度慢，而且在磁盘控制器被指示到磁盘检索出所需要的数据之前，还要采取额外的步骤。然而，通过让处理器向高速缓存控制器发调用，再由后者确定磁盘是否实际读出，便可以显著提高性能。一些高速缓存控制器厂商声称，使用智能高速缓存控制器，可使性能大为改善。高速缓存控制器已经在大型计算机系统上使用多年，将它们向微机上移植，可以看成合乎逻辑的下一步。

1.4 输入/输出系统

1.4.1 课文 A
键盘

若熟悉打字机，你会发现计算机键盘的布局与其非常相似。使用键盘可以进行如下工作：
- 输入信息
- 用数字小键盘输入数字

- 请求特殊功能
- 用键组合执行系统功能
- 在屏幕上移动

键盘有字母键、标点符号键和一个空格键。它还有功能键、数字键和箭头键。如何使用这些键，取决于计算机上所安装的软件。软件所带的文档包含关于特殊键功能的信息。你也许注意到了计算机键盘的触感（响应）与打字机的有所不同。计算机的键盘非常敏感，只需轻轻触摸便可输入。当保持按下一个字符键时，该字符将持续输入。这就称为计算机键盘的复击效应。

显示器

显示器也许是最重要的输出设备之一。计算机只能用它们来显示有趣的结果和神奇生动的画面。显示器也是人机对话的最好窗口，所以很多用户选择显示器时非常小心。选择显示器时应该注意哪些参数和指标呢？我们在这里给出一些参数供你参考。

像素距离：两个图像像素水平方向的距离称为像素距离。目前大多数 PC 显示器的流行像素距离是 0.28mm。这个值越小，屏幕越清晰。

视频带宽：它是显示器技术中一个很重要的概念，它关系到显示器的最高工作频率。它的范围从几十兆赫到几百兆赫。

分辨率：它是显示器的另一个重要参数，其值越高，屏幕上的图像越清晰。分辨率表示一个屏幕上全部图像像素的总和。

扫描方式：显像管中的电子枪的扫描方式有两种：隔行和非隔行。在隔行方式中，电子束首先扫描奇数行中的像素，第二次再扫描偶数行中的像素。一帧画面的更新需扫描两次。在非隔行的方式中，电子束一次完成扫描全部像素。在这种工作方式中，显示器工作得更好且图像清晰、不闪烁。

鼠标

鼠标与系统之间的接口可以采用如下两种方式之一：通过鼠标的移动产生一系列脉冲（使用 LED 和检测器产生脉冲）或者利用鼠标使计数器增值或减值。处理器能定期读取这些计数器中的值或对脉冲计数，并且确定自上一次检测点处算起，鼠标移动了多远。然后系统把光标移到屏幕的恰当位置。这种移动看起来是很平滑的，因为鼠标的移动速度与处理器读取鼠标的位置及移动屏幕上光标的速度相比是很低的。

大部分鼠标还装有一个或多个按键，并且当按下按键时，系统必须能够及时检测到。依靠对按键状态的监测，系统也就能区分单击按键与按着按键拖曳之间的差别。当然，计数器与按键实际位置之间的转换及屏幕上的变化都是由软件控制的。这也正说明了为什么鼠标在屏幕上的移动速度及鼠标按键是单击还是双击的识别速度能够由用户来设定。同样，用软件描述鼠标位置，也意味着当鼠标沿一个方向长距离移动时，光标不会完全跳离屏幕。

光盘

光盘是一种用激光对数据进行编码以便检索的盘片。它提供的信息密度远远超过现行的磁性大容量存储设备的范围。类似的设备已经在市场上出现了几年的时间，以激光视盘和音频致密光盘（CD）的形式供用户使用。这些激光视盘是模拟的，即光盘包含了像唱片那样的一条螺旋记录道。用于计算机的光盘是数字式的，像磁盘一样，信息存储在同心道上。目前，有三种类型的光盘技术竞争于大容量存储器市场，它们是只读光盘、一次写入型光盘和可擦除光盘。

与常规的磁盘不同，只读光盘不可写入，所以与只读存储器有等价的功能。最流行的只读光盘的版本采用的技术与已流行的用于音频唱片的 CD 技术相同。这种技术是数字式的，并且基于单面能存储 540MB 的 4¾ 英寸的光盘。这种设备称为致密光盘只读存储器（CD-ROMs）。

一次写入型光盘（也称一次写入多次读出光盘，或 WORM）是由用户记录的空白盘。为了写入数据，激光的强大光束将覆盖在光盘的表层并烧结出小斑点或凹点。一旦烧出，这些斑点就不能被擦除。为了检索数据，使用不太强的激光去读斑点的模式，并且把该模式转变为可以在电视机上播放的视听信号。一次写入型光盘可用来代替微缩胶片存储器。因为光盘具有存储图像和声音的能力，所以它们的用途是多方面的。任何能够被数字化的事物，比如文件、图像、照片、图画和音乐都可以记录和存储在光盘上。

可擦除光盘使用激光从光盘读出或向光盘写入信息，不过光盘的表面也使用磁性材料和磁性写磁头以获得可擦性。为了向光盘写入，激光束在光盘上加热成小点，接着提供一个磁场以改变点的磁性。可擦除光盘既提供了与非擦除光盘相同的存储能力，又具有同温彻斯特系统那样的常规磁盘的重复使用能力。

1.4.2　课文 B

总线是共享的通信链路，它用一束通信线来连接多个子系统。总线结构的两个最主要的优点是：灵活性和廉价性。通过定义一种连接方式，可以很容易地增加新设备，甚至也可将外围设备在两个使用同类总线的计算机系统间移动。此外，总线能被有效地使用，因为一条总线可以多种方式共享。总线的最大缺点是它产生通信瓶颈，可能限制最大 I/O 流量。当 I/O 必须经过一条总线时，总线带宽限制了最大 I/O 流量。

总线设计如此困难的一个原因是，总线的速度在很大程度上受物理因素限制：总线长度和设备数量。这些物理因素使我们不能以任意高速使用总线。在这些限制下，尽管我们可用许多技术来提高总线的性能，但是，这些技术也会对其他性能产生不利影响。例如，为获得较快的 I/O 响应时间，我们必须用流水线化通信路径来最小化总线延时。另一方面，为维持高速 I/O 数据速率，我们必须最大化总线宽度。使用更大的缓冲器和用更大的数据块通信能增加总线宽度，但两者都增加总线的延时。显然，这两个目标，即低延时和高带宽，会导致设计需求的冲突。最后，要求设备支持具有不同延时和数据传输速率的各种各样的需求，也使总线设计面临挑战。

一条总线一般包括一束控制线和一束数据线。控制线被用于标记请求和应答，并且指出数据线上信息的种类。总线的数据线在源地和目的地间传送信息。这种信息可以包括数据、复杂命令和地址。比如，磁盘想要从某个扇区往内存中写数据，数据线就用于指明内存中的哪个地址用于存放数据，同时将数据从磁盘输入内存，控制线用于指出传输的每一时刻数据线上都包括哪种类型的信息。一些总线有两束信号线，在一个总线传输中分别传输数据和地址。在任何一种情况下，控制线都用于指出总线包括什么并且执行总线协议。

每个 I/O 设备与计算机系统的地址总线、数据总线和控制总线相连接，它们都包括 I/O 接口电路。与总线交互的实际上正是这一电路，同时，它与实际的 I/O 设备交互来传输数据。

对于一个输入设备的一般接口电路，从输入设备来的数据传送到三态缓冲器。当地址总线和控制总线上的值正确时，缓冲器设为有效，数据传到数据总线上，然后，CPU 可以读取数据。当条件不正确时，逻辑块不会使缓冲器有效，它们是三态的，不把数据传到总线上。这一设计的关键在于使能逻辑。正如每个存储单元都有一个唯一的地址一样，每个 I/O 设备也有一个唯一的地址。除非从地址总线得到了正确的地址，否则使能逻辑不使缓冲器有效。同时，它还必须从控制总线上得到正确的控制信号。

输出设备接口电路的设计与输入设备的设计有所不同，寄存器代替了三态缓冲器。输入设备中使用了三态缓冲器是为了确保在任何时刻都只有一个设备向总线写数据。而输出设备是从总线读取数据，不是写数据，因此不需要缓冲器。数据对于所有的输出设备都可获得，但只有具有正确地址的设备才会读取它。

Unit 2　Software Knowledge

2.1　Data Structure

2.1.1　Text A

A data structure is a data type whose values are composed of component elements that are related by some structure.[1] It has a set of operations on its values. In addition, there may be operations that act on its component elements. Thus we see that a structured data type can have operations defined on its component values, as well as on the component elements of those values.

Data type

The essence of a data type is that it attempts to identify qualities common to a group of individuals or objects that distinguish it as an identifiable class or kind.[2] If we provide a set of possible data values and a set of operations that act on the values, we can think of the combination as a data type.

Let us look at two classes of data types. We will call any data type whose values we choose to consider atomic an atomic data type. Often we choose to consider integers to be atomic. We are then only concerned with the single quantity that a value represents, not with the fact that an integer is a set of digits in some number system. Integer is a common atomic data type found in most programming languages and in most computer architectures.

We will call any data type whose values are composed of component elements that are related by some structure a structured data type, or data structure. In other words, the values of these data types are decomposable, and we must therefore be aware of their internal construction. There are two essential ingredients to any object that can be decomposed—it must have component elements and it must have structure, the rules for relating or fitting the elements together.

Object-oriented data structure

Object-oriented software development is a contemporary approach to the design of reliable and robust software. The complexity of the implementation of software system is a combination of the complexity of the representations of information and the complexity of the algorithms that manipulate the representations. Data structure is the study of methods of representing objects, the safe, reliable encapsulation of structure, the development of algorithms that use these representations, and the measurement of both the time and space complexity of the resulting systems. The object-oriented approach emphasizes the role of objects, along with their attributes and operations, that form the nucleus of the solution.

From the point of view of deciding which data structure should represent that attributes of objects in a specific class, the emphasis that the object-oriented approach places on abstraction is very important to the software development process. Abstraction means hiding unnecessary details. Procedural abstraction, or algorithmic abstraction, is hiding of algorithmic details, which allows the algorithm to be seen or described, at various levels of detail. Building subprograms so that the names of the subprograms describe what the subprograms do and the code inside subprograms shows how the processes are accomplished is an illustration of abstraction in action.

Similarly, data abstraction is the hiding of representational details. An obvious example of this is the building of data types by combining together other data types, each of which describes a piece, or attribute, of a more complex object type. An object-oriented approach to data structures brings together both data abstraction and procedural abstraction through the packaging of the representations of classes of objects.

Once an appropriate abstraction is selected, there may be several choices for representing the data structure. In many cases there is at least one static representation and at least one dynamic representation. The typical tradeoff between static and dynamic representations is between a bounded or unbounded representation versus the added storage and time requirements associated with some unbounded representations.

After an abstraction and representation are chosen, there are competing methods to encapsulate data structures.[3] The choice of an encapsulation is another tradeoff, between how the structure is made available to the user and how the user's instantiating objects may be manipulated by the package. The encapsulations have an effect on the integrity of the representation, and time and space requirements associated with the encapsulation. Once specified, one or more competing methods of representation may be carried out, and the structure, its representations and its encapsulation may be evaluated relative to the problem being solved. The time and space requirements of each method must be measured against system requirements and constraints.

Object-oriented programming differs from procedural programming because it uses objects as data structure.[4] The structured data and its related operations could be encapsulated in a single object which may be reused and easily upgraded, augmented, replaced. So it directly reduces the cost of maintenance and the timing and extendibility of new system.

Key Words

abstract	抽象，摘要
atomic	原子的，极微的
contemporary	当代的，现代的

decomposable	可分解的
encapsulation	封装，压缩
essence	本质，实质，精华，要素
extendibility	可扩充性，可扩展性，可延伸性
illustration	说明，例证
ingredient	组成，成分
quantity	数量，定量
representational	代表的，表现的
unbounded	无限的，无边际的，极大的

Notes

[1] A data structure is a data type whose values are composed of component elements that are related by some structure.

说明：由 whose 引导的限定性定语从句修饰 a data type，that 引导的限定性定语从句修饰 component elements。

译文：数据结构是一种数据类型，其值是由与某些结构有关的组成元素所构成的。

[2] The essence of a data type is that it attempts to identify qualities common to a group of individuals or objects that distinguish it as an identifiable class or kind.

说明：本句由两个复合句构成，均由 that 引导。第一个 that 引导表语从句；第二个 that 引导限定性定语从句，修饰 qualities，it 代表 a group of individuals or objects。

译文：数据类型的本质是标识一组个体或目标所共有的特性，这些特性把该组个体作为可识别的种类。

[3] After an abstraction and representation are chosen, there are competing methods to encapsulate data structures.

说明：句中的"After"引导的是时间状语从句。

译文：在选择了抽象和表示后，就有各种不同的方法来封装数据结构。

[4] Object-oriented programming differs from procedural programming because it uses objects as data structure.

说明：句中的"because"引导的是原因状语从句。

译文：由于面向对象程序设计把对象当作数据结构，所以它不同于过程程序设计。

2.1.2 Text B

Stack and Queue

A stack is a data type whose major attributes are determined by the rules governing the insertion and deletion of its elements. The only element that can be deleted or removed is the one that was inserted most recently. Such a structure is said to have a last-in/first-out (LIFO) behavior, or protocol. The simplicity of the data type stack belies its importance. Many computer systems have stacks built into their circuitry and have machine-level instructions to operate the hardware stack. The sequencing of calls to and returns from subroutines follows a stack protocol. Arithmetic expressions are often evaluated by a sequence of operations on a stack. Many handheld calculators use a stack mode of operation. In studying computer science, you can expect to see many examples of stacks.

Queues occur frequently in everyday life and are therefore familiar to us. The line of people waiting for service at a bank or for tickets at a movie theater and the line of autos at a traffic light are examples of queues. The main feature of queues is that they follow a first-come/first-served rule. Contrary to a stack, in which the latest element inserted is the first removed or served, in queues the earliest element inserted is the first served. In social settings, the rule appeals to our sense of equality and fairness.

There are many applications of the first-in/first-out (FIFO) protocol of queues in computing. For example, the line of input/output (I/O) requests waiting for access to a disk drive in a multi-user time-sharing system might be a queue. The line of computing jobs waiting to be run on a computer system might also be a queue. The jobs and I/O requests are serviced in order of their arrival, that is, the first in is the first out. There is a second kind of queue that is important. An everyday example can be seen in an emergency room of a hospital. In large emergencies it is common to first treat the worst injured patients who are likely to survive.

In computer systems, events that demand the attention of the computer are often handled according to a most-important-event/first-served, or highest-priority in/first-out (HPIFO), rule. Such queues are called priority queue, in this type of queue service is not in order of time of arrival but rather in order of some measure of priority.

Function calls

When a call is made to a new function, all the variables local to the calling routine need to be saved by the system, since otherwise the new function will overwrite the calling routine's variables. Furthermore, the current location in the routine must be saved so that the new function knows where to go after it is done. The variables have generally been assigned by the compiler to machine registers, and there are certain to be conflicts, especially if recursion is involved.

When there is a function call, all the important information that needs to be saved, such as register values and the return address, is saved "on a piece of paper" in an abstract way and put at the top of a pile. Then the control is transferred to the new function, which is free to replace the registers with its values. If it makes other function calls, it follows the same procedure. When the function wants to return, it looks at the "paper" at the top of the pile and restores all the registers. It then makes the return jump.

Clearly, all of this work can be done using a stack, and that is exactly what happens in virtually every programming language that implements recursion. The information saved is called either an activation record or stack frame. The stack in a real computer frequently grows from the high end of your memory partition downwards, and on many systems there is no checking for overflow. There is always the possibility that you will run out of stack space by having too many simultaneously active functions. Needless to say, running out of stack space is always a fatal error.

In languages and systems that do not check for stack overflow, your program will crash without an explicit explanation. On these systems, strange things may happen when your stack gets too big, because your stack will run into part of your program. It could be the main program, or it could be part of your data, especially if you have a big array. If it runs into your program, your program will be corrupted; you will have nonsense instructions and will crash as soon as they are executed. If the stack runs into your data, what is likely to happen is that when you write something into your data, it will destroy stack information—probably the return address—and your program will attempt to return to some weird address and crash.

Key Words

appeal	要求，呼吁，倾向于
assign	分派，指派，赋值
conflict	冲突，矛盾
deletion	删除
emergency	紧急，应急
injured	受伤的，受损害的
insertion	插入
overflow	溢出
pile	堆
priority	优先权，优先级
recursion	递归，循环
routine	例行的，常规的，程序
stack	栈，堆栈
survive	生存

2.1.3 Exercises

1. Translate the following phrases into English

(1) 内部结构
(2) 数据类型
(3) 结构数据类型
(4) 算术表达式
(5) 函数调用

2. Translate the following phrases into Chinese

(1) last in/first out
(2) atomic data type

(3) stack space

(4) data structure

(5) return address

3. Identify the following to be True or False according to the text

(1)The sequencing of calls to and returns from subroutines follows a stack protocol.

(2)Both the stack and the queue have the same behaviors.

(3)The priority queues can use HPIFO rule.

(4)We call the data type whose values are not composed of component elements a structured data type.

(5)In languages and systems that do not check for stack overflow, your program will crash without an explicit explanation.

4. Reading Comprehension

(1)The operations of a structured data type might act on _____.

a. component values

a. component elements

b. either component values or component elements

c. neither component values nor component elements

(2)We will call any data type whose values we choose to consider atomic_____.

a. an atomic data type

b. a structured data type

c. data type

d. local data

(3)The main feature of queues is that they follow a _____ rule.

a. last-in/first-out

b. first-in/last-out

c. first-come/last-served

d. first-in/first-out

(4)The main feature of stacks is that they follow a _____ rule.

a. last-in/last-out

b. first-in/last-out

c. first-in/first-out

d. highest-priority in/first-out

2.2 System Software

2.2.1 Text A

An operating system is a program, which acts as an interface between a user of a computer and the computer hardware.[1] The purpose of an operating system is to provide an environment in which a user may execute programs. In general, however, there is no completely definition of an operating system. Operating systems exist because they are a reasonable way to solve the problem of creating a usable computing system. The fundamental goal of computer systems is to execute user programs and solve user problems. Towards this goal computer hardware is constructed. Since bare hardware alone is not very easy to use, application programs are developed. These various programs require certain common operations, such as controlling the I/O devices. The common functions of controlling and allocating resources are then brought together into one piece of software: the operating system.

The first operating systems were developed by manufactures for the computers in their product line. When the manufacturers came out with another computer or model, they often produced an improved and different system; this meant that users who wanted to switch computers, either from one vendor to another or to a different model from the same vendor, would have to convert their existing programs to run under the new operating system. Today, however, the trend is away from operating system limited to a specific model and toward operating systems that will run on any model by a particular manufacturer.

There are many important reasons for learning operating system; the most notable are:

♦ The user must interact with the operating system in order to accomplish task since it is his primary interface with the computer;

♦ The selection of the operating system and its options is a major decision for most computer installations;

♦ Many concepts and techniques found in operating systems have general applicability in other applications;

♦ For special-purpose usage you may have to design your own operating system or modify on existing one.

An operating system is similar to a government. Its hardware, software, and data provide the basic resource of a computer system. The operating system provides the means for the proper use of these resources in the operation of the computer system. Like government, the operating system simply provides an environment within which other programs can do useful work.

We can view an operating system as a resource allocates. A computer system has many resources which may be required to solve a problem: CPU time, memory space, file storage space, input/output devices, and so on. The operating system acts as the manager of these resources and allocates them to specific programs and users as necessary for their tasks. Since there may be many, possibly conflicting, requests for resources, the operating system must decide which requests are allocated resources to operate the computer system fairly and efficiently.[2]

Operating systems are either single-tasking or multitasking. The more primitive single-tasking operating systems can run only one process at a time. For instance, when the computer is printing a document, it cannot start another process or respond to new commands until the printing is completed.

All modern operating systems are multitasking and can run several processes simultaneously. In most computers there is only one CPU, so a multitasking operating system creates the illusion of several processes running simultaneously on the CPU. The most common mechanism used to create this illusion is time slice multitasking, whereby each process is run individually for a fixed period of time.[3] If the process is not completed within the allotted time, it is suspended and another process is run. This exchanging of processes is called context switching. The operating system performs the "bookkeeping" that preserves the state of a suspended process. It also has a mechanism, called a scheduler, that determines which process will be run next. The scheduler runs short processes quickly to minimize perceptible delay. The processes appear to run simultaneously because the user's sense of time is much slower than the processing speed of the computer.

A very important responsibility of any operating system is the scheduling of jobs to be handled by a computer system.[4] This is one of the main tasks of the job management function. The operating system sets up the order in which programs are processed, and defines the sequence in which particular jobs are executed. The term job queue is often used to describe the series of jobs awaiting execution. The operating system weighs a variety of factors in creating the job queue. These include which jobs are currently being processed, the system's resources being used, which resources will be needed to handle upcoming programs, the priority of the job compared to other tasks, and any special processing requirements to which the system must respond. The operational software must be able to assess these factors and control the order in which jobs are processed.

Key Words

applicability	应用性，适用性
bookkeeping	簿记，记账
environment	环境
fundamental	基本的，原则
government	政府，政体
handle	操作，操纵
illusion	幻影，错觉
installation	安装
mechanism	机制
perceptible	可察觉的，看得见的
preserve	保存，维持
primitive	原始的，基本的
proper	适当的，正式的
reasonable	合理的，适当的

resource	资源
responsibility	职责，责任
suspend	暂停，挂起
vendor	供应商，卖主

Notes

[1] An operating system is a program, which acts as an interface between a user of a computer and the computer hardware.

说明：由"which"引导非限定性定语从句，修饰"program"。

译文：操作系统是一种程序，它是用户与计算机硬件之间的接口。

[2] Since there may be many, possibly conflicting, requests for resources, the operating system must decide which requests are allocated resources to operate the computer system fairly and efficiently.

说明：本句由"Since"引导原因状语从句，主语是"the operating system"，"which requests are allocated…"是宾语从句。

译文：因为有许多资源请求方面可能存在冲突，操作系统必须决定给哪些请求分配资源以便计算机系统能合理而有效地运行。

[3] The most common mechanism used to create this illusion is time slice multitasking, whereby each process is run individually for a fixed period of time.

说明：过去分词短语 used to create this illusion 作为定语，修饰 mechanism；由 whereby 引导的是非限制性定语从句。

译文：产生这种错觉的最常用机制是时间分割多任务处理，它是以每个进程各自运行固定的一段时间的方式来实现的。

[4] A very important responsibility of any operational software is the scheduling of jobs to be handled by a computer system.

说明：这里的"to be handled by a computer system"是不定式短语作为定语。

译文：操作软件的一个非常重要的职责是调度计算机系统将要处理的作业。

2.2.2 Text B

A compiler is a program which takes text in high-level language and converts it into equivalent text in a low-level assembly or machine language. Let's overview the seven major components common to most existing compilers and to examine their interconnections. The seven components are as follow:

- Scanner
- Parser
- Intermediate code generator
- Semantic processor
- Optimizer

- Code generator
- Tables

Not all compilers have each of these modules as physically independent entities. For example, the scanner may be integrated into the parser, or semantic processor may be integrated into the parser, and so on. Conceptually, however it is easier to decompose any compiler into these seven components.

The scanner in other books is called lexical analyzer. When a program is first input into a compiler, it is just one long stream of characters, perhaps broken into lines or records reflecting the input medium. The scanner converts this external view of the source program into an internal format more suited for further manipulation by the remainder of the compiler.

The scanner has several roles:
- Identify the basic lexical units of the program, which are called tokens;
- Remove extraneous blanks, carriage returns and other characteristics of the input medium;
- Remove comments;
- Report errors which the scanner discovers.

Typically, a scanner will make one pass over the text in its original form, carrying out its four tasks as this pass proceeds and output the program in some internal format, a token at a time, to the parser upon request or as one large file of tokens. Usually, the scanner examines the text character by character.

Each programming language has its own set of grammar rules characterizing the correct form of programs in the language. The parser or syntactic analyzer accepts the output of the scanner; i.e., token and verifies that the source program satisfies the grammatical rules of the language being compiles. The tree output by the parser is transformed into a "program" of sorts written in an intermediate code, which is closer in form to assembly language than source text and yet is in a form which makes further manipulation easier than if actual assembly (or machine) code were emitted.

Intermediate code can be converted directly into the language by the code generator. However, it is common to insert another component between the semantic processor and the code generator; namely, the optimizer (code optimization). If the code generator transforms intermediate level text into assembly language in a straightforward manner, the generated object codes probably not as efficient with respect to execution time and storage space as it might be. If a compiler is more than superficially concerned with producing efficient code, as most commercial compilers must be, it will include a module specifically designed to improve some combination of time and space characteristics of the code. The optimizer modifies the code which it is given into a more efficient version.

We classify optimization techniques into two categories: those which are performed on the source program (its internal form) and which are therefore independent of the object language, and those which are performed on the object program level.

The code generator takes the intermediate code it receives the optimizer and produces assembly or machine language code (the object program); obviously code generation is highly machine-dependent. Hence, whenever the object machine is altered, the code generator must be extensively revised. On the other hand, the other components are somewhat more insensitive to the machine for which code is being generated; however, none of the compiler modules is, in general, totally immune to a change in the target machine.

Key Words

conceptually	概念地
equivalent	相等的，相当的，等价的
extraneous	非必要的，无关的
grammatical	语法上的，符合语法规则的
immune	不受影响的
insensitive	敏感的，灵敏的
intermediate	中间的，中间体
lexical	词汇的，词法的
optimizer	优化器，优化程序
parser	语法分析器，语法分析程序
reflect	反射，反映，映出
scanner	扫描程序，扫描器，扫描设备
semantic	语义的，语义学的
superficially	表面地，肤浅地
syntactic	句法的

2.2.3 Exercises

1. Translate the following phrases into English
(1)文件存储空间
(2)操作系统
(3)代码生成器
(4)词法分析器
(5)资源管理器

2. Translate the following phrases into Chinese
(1)allocate resource
(2)single-task
(3)syntactic analyzer
(4)code optimization
(5)semantic processor

3. Identify the following to be True or False according to the text

(1) Without an operating system, we could not execute a user program.

(2) Operating systems can only be used in multitask systems.

(3) The operating system sets up the order in which programs are processed, and defines the sequence in which particular jobs are executed.

(4) The scanner may be integrated into the parser.

(5) Intermediate code can't be converted directly into the language by the code generator.

4. Reading Comprehension

(1) The _____ serves as an interface between hardware and software.

a. system

b. application program

c. operating system

d. control unit

(2) The term _____ is often used to describe the series of jobs awaiting execution.

a. file queue

b. task queue

c. job queue

d. process queue

(3) Which module does not belong to a compiler? _____

a. operating system

b. scanner

c. optimizer

d. tables

(4) Each programming language has its own set of _____ characterizing the correct form of programs in the language.

a. instructions

b. codes

c. comments

d. grammar rules

2.3 C Language and C++ Language

2.3.1 Text A

C is a general-purpose, structured programming language. Its instructions consist of terms that resemble algebraic expressions, augmented by certain English keywords such as if, else, for, do and while. In this respect C resembles other high-level structured programming languages. C also contains certain additional features however, that allow it to be used at a lower level, thus bridging the gap between machine language and the more conventional high-level languages. This flexibility allows C to be used for systems programming (e.g., for writing operating systems) as well as for applications programming (e.g., for writing a program to solve a complicated system of mathematical equations, or for writing a program to bill customers).

C was developed in the early 1970s. C might best be described as a "medium-level language". Like a true high-level language, there is a one-to-many relationship between a C statement and the machine language instructions it is complied into. Thus, a language like C gives you far more programming leverage than a low-level assembly language. However, compared to most high-level language, C has a very small set of constructs.[1] In addition, unlike most high-level language, C lets you easily do chores (such as bit and pointer manipulation) additionally performed by assembly language.

Structured language

Although the term block-structured language does not strictly apply to C in an academic sense, C is informally part of that language group.[2] The distinguishing feature of block-structured language is that the compartmentalization of code and data. This means that a language can section off and hide from the rest of the program all information and instructions that are necessary to perform a specific task. Generally, compartmentalization is achieved by subroutines with local, or temporary, variables. In this way, you can write subroutines so that the events that occur within them will cause no side effects in other parts of the program. Excessive use of global variables, which are known throughout the entire program, may allow bugs, or unwanted side effects, to creep into a program. In C, all subroutines are discrete functions.

Functions are the building blocks of C, in which all program activity occurs. They allow you to define and code specific tasks in a program separately. After debugging a function that uses only local variables, you can rely on it to work properly in various situations without creating side effects in other parts of your program. All variables that are declared in that function will be known only to that function.

A C program consists of a series of functions. Program execution must begin with a function called main(). Other functions included in the program are named by the programmer. When the name of a function appears as a statement in a program, program execution transfers to that function. After the called function has been completely executed, a result is made to the calling function. If one function calls another function, the second function is said to be nested inside the first. In many cases, when a function is called, information is passed to it. This information is included within the parentheses after the function name. The called function may also return a single value to the calling function.

Characteristics

C is characterized by the ability to write very concise source programs, due in part to the large number of operators included within the language. It has a relatively small instruction set, though actual implementations include extensive library functions which enhance the basic instructions. Furthermore, the language encourages users to write additional library functions of their own. Thus, the features and capabilities of C language can easily be extended by the user.

C compilers are commonly available for computers of all sizes, and C interpreters are becoming increasingly common.[3] The compilers are usually compact, and they generate object programs that are small and highly efficient when compared with programs compiled from other high-level languages, the interpreters are less efficient, though they are easier to use when developing a new program. Many programmers begin with an interpreter, and then switch to a compiler once the program has been debugged (once all of the programming errors have been removed).

Another important characteristic of C is that its programs are highly portable, even more so than with other high-level languages. The reason for this is that C relegates most computer-dependent features to its library functions. Thus, every version of C is accompanied by its own set of library functions, which are written for the particular characteristics of the host computer.[4] These library functions are relatively standardized, however, and each individual library function is generally accessed in the same manner from one version of C to another. Therefore, most C programs can be processed on many different computers with little or no alteration.

Key Words

algebraic	代数的，关于代数的
assembly	汇编，装配，集合
augment	增强，加强
compartmentalization	划分，分门别类
compiler	编译器，编译程序
creep	蔓延，爬行
debug	调试
flexibility	灵活性，适应性
leverage	杠杆作用
nest	嵌套，一套物件
portable	手提的，轻便的
resemble	相似，一致，类似
standardized	标准的，定型的
strictly	严格地，完全地

Notes

[1] However, compared to most high-level language, C has a very small set of constructs.
说明：本句中的"compared to"的含义是"将……与……比较"。

译文：与大多数高级语言相比，C 语言有一个很小的结构集。

[2] Although the term block-structured language does not strictly apply to C in an academic sense, C is informally part of that language group.

说明：本句中"Although"引导的是让步状语从句。

译文：尽管块结构语言这个术语从学术上考虑并不能严格地应用于 C 语言，但 C 语言是那个语言族的非正式成员。

[3] C compilers are commonly available for computers of all sizes, and C interpreters are becoming increasingly common.

说明：本句是一个并列句，"C compilers"和"C interpreters"是主语。

译文：C 语言的编译程序普遍适用于各种容量的计算机，并且 C 语言的解释程序正变得越来越普通。

[4] Thus, every version of C is accompanied by its own set of library functions, which are written for the particular characteristics of the host computer.

说明：本句中的"which are written for…"做非限定性定语从句，修饰"library functions"。

译文：因此，每个版本的 C 语言都伴有它自己的库函数集，这些库函数集是按主机的特点而编写的。

2.3.2　Text B

C++ is a general-purpose programming language with high-level and low-level capabilities. It is a statically typed, free-form, multi-paradigm, usually compiled language supporting procedural programming, data abstraction, object-oriented programming, and generic programming. C++ is an enhanced version of the C. C++ includes everything that is part of C and adds support for object-oriented programming. C is a procedure-oriented programming language. C++ contains many improvements and features. The basic syntax and semantics of C and C++ are the same. If you are familiar with C, you can program in C++ immediately. C++ has the same types, operators, and other facilities defined in C that usually correspond directly to computer architecture.

C++ fully supports object-oriented programming, including the four pillars of object-oriented development: encapsulation, data hiding, inheritance, and polymorphism. While it is true that C++ is a superset of C, and that virtually any legal C program is a legal C++ program, the leap from C to C++ is very significant. C++ benefited from its relationship to C for many years, as C programmers could ease into their use of C++. To really get the full benefit of C++, however, many programmers found they had to unlearn much of what they knew and learn a whole new way of conceptualizing and solving programming problems.

Object-oriented programming is a programming technique that allows you to view concepts as a variety of objects. By using objects, you can represent the tasks that are to be performed, their interaction, and any given conditions that must be observed. A data structure often forms the basis of an object; thus, in C or C++, the struct type can form an elementary object.

Communicating with objects can be done through the use of messages. Using messages is similar to calling a function in a procedure-oriented program. When an object receives a message, methods contained within the object respond. Methods are similar to the functions of procedure-oriented programming. However, methods are part of an object.

C++ supports the properties of encapsulation and data hiding through the creation of user-defined types, called classes. Once created, a well-defined class acts as a fully encapsulated entity—it is used as a whole unit. The actual inner workings of the class should be hidden. Users of a well-defined class do not need to know how the class works; they just need to know how to use it.

The C++ class is an extension of the C structure. Because the only difference between a structure and a class is that structure members have public access by default and a class member has private access by default, you can use the keywords class or struct to define equivalent classes. The C++ class is an extension of the C and C++ struct type and forms the required abstract data type for object-oriented programming. The class can contain closely related items that share attributes. Stated more formally, an object is simply an instance of a class. Ultimately, there should emerge class libraries containing many object types. You could use instances of those object types to piece together program code.

Typically, an object's description is part of a C++ class and includes a description of the object's internal structure, how the object relates with other objects, and some form of protection that isolates the functional details of the object from outside the class. The C++ class structure does all of this.

In a C++ class, you control functional details of the object by using private, public, or protected descriptors. In object-oriented programming, the public section is typically used for the interface information (methods) that makes the class reusable across applications. If data or methods are contained in the public section, they are available outside the class. The private section of a class limits the availability of data or methods to the class itself. A protected section containing data or methods is limited to the class and any derived subclasses.

Key Words

encapsulation	封装，包装
facility	方便，简便，机构
inheritance	继承，承受
interaction	互动，交互，相互作用
leap	跳过，跃过
pillar	柱，台柱，核心
polymorphism	多态性
section	部分，部件
semantics	语义学
statically	静止地，静态地

subclass	子类，子集
superset	超集
syntax	语法
ultimately	最终

2.3.3 Exercises

1. Translate the following phrases into English

(1)高级语言

(2)源代码

(3)库函数

(4)面向过程的编程语言

(5)数据隐藏

2. Translate the following phrases into Chinese

(1) structured programming language

(2) machine language

(3) assembly language

(4) object-oriented program

(5) functional detail

3. Identify the following to be True or False according to the text

(1)There is a one-to-one relationship between a C statement and a machine language instruction.

(2)A C program must begin with a function called main().

(3)Most C programs can be processed on many different computers with little or no alteration.

(4)Object-oriented programming is a programming technique that allows you to view concepts as a variety of objects.

(5)The C++ class is not an extension of the C language structure.

4. Reading Comprehension

(1)C++ fully supports object-oriented programming, including the four pillars of object-oriented development: _____.

a. data type, data hiding, inheritance, and polymorphism

b. encapsulation, data hiding, inheritance, and data type

c. encapsulation, data type, inheritance, and polymorphism

d. encapsulation, data hiding, inheritance, and polymorphism

(2)When the name of a function appears as a statement in a program, program execution transfers to that _____.

a. name

b. data

c. function

d. main()

(3)C was developed in the early _____.

a. 1970s

b. 1950s

c. 1980s

d. 1860s

(4)After the called function has been completely executed, a result is made to the _____.

a. called function

b. main function

c. function

d. calling function

2.4 Java and Object-Oriented Programming

2.4.1 Text A

Java has become enormously popular. Java's rapid rise and wide acceptance can be traced to its design and programming features, particularly its promise that you can write a program once and run it anywhere. As stated in the Java-language white paper by Sun, Java is simple, object-oriented, distributed, interpreted, robust, secure, architecture-neutral, portable, high-performance, multithreaded, and dynamic. In brief, Java environment can be used for developing the application software that can be operated on any computing platform. It is a kind of basic technology with compact structure in fact, and its overall influence on World Wide Web and commerce can correctly be compared with the impact of electronic form on PC.

Java is a bit easier than the popular object-oriented programming language C++, which was the dominant software-development language before Java. Java is partially modeled on C++, but greatly simplified and improved. For instance, pointers and multiple inheritance often make programming complicated, Java replaces the multiple inheritance in C++ with a simple language construct called an interface, and eliminates pointers.

Java uses automatic memory allocation and garbage collection, whereas C++ requires the programmer to allocate memory and collect garbage. Also, the number of language constructs is small for such a powerful language. The clean syntax makes Java programs easy to write and read.

- Java is object-oriented

Although many object-oriented languages began strictly as procedural language, Java was designed from the start to be object-oriented. Object-oriented programming (OOP) is popular programming approach that is replacing traditional procedural programming techniques.

Software systems developed using procedural programming languages are based on the paradigm of procedures. Object-oriented programming models the real world in terms of objects. Everything in the world can be modeled as an object. A circle is an object, a person is an object, and a Windows icon is an object. Even a loan can be perceived as an object. A Java program is object-oriented because programming in Java is centered on creating objects, manipulating objects, and making objects work together.

One of the central issues in software development is how to reuse code. Object-oriented programming provides great flexibility, modularity, clarity, and reusability through encapsulation, inheritance, and polymorphism. For years, object-oriented technology was perceived as elitist, requiring a substantial investment in training and infrastructure. Java has helped object-oriented technology enter the mainstream of computing. Java's simple, clean syntax makes programs easy to write and read. Java programs are quite expressive in terms of designing and developing applications.

- Java is distributed

Distributed computing involves several computers working together on a network. Java is designed to make distributed computing easy. Since networking capability is inherently integrated into Java, writing network programs is like sending and receiving data to and from a file. [1] If you download a Java applet (a special kind of program) and run it on your computer, it will not damage your system because Java implements several security mechanisms to protect your system against harm caused by stray programs.

- Java is interpreted

You need an interpreter to run Java programs. The programs are compiled into the Java virtual machine code called byte-code. The byte-code is machine-independent and can run on any machine that has a Java interpreter, which is part of the Java virtual machine.

Most compilers, including C++ compilers, translate programs in a high-level language to machine code. The code can only run on the native machine. If you run the program on other machine, it has to be recompiled on the native machine. [2] For instance, if you compile a C++ program in Windows, the executable code generated by the compiler can only run on the Windows platform. With Java, you compile the source code once, and the byte-code generated by a Java compiler can run on any platform with a Java interpreter. The Java interpreter translates the byte-code into the machine language of the target machine.

- Java is robust

Robust means reliable. No programming language can ensure complete reliability. Java puts a lot of emphasis on early checking for possible errors, because Java compilers can detect many problems that would first show up at execution time in other language. Java has eliminated certain types of error-prone programming constructs found in other languages. It does not support pointers, for example, thereby eliminating the possibility of overwriting memory and corrupting data.

Java has a runtime exception-handling feature to provide programming support for robustness. Java forces the programmer to write the code to deal with exceptions. [3] Java can catch and respond to an exceptional situation so that the program can continue its normal execution and terminate gracefully when a runtime error occurs. [4]

- Java is architecture-neutral

Java is interpreted. This feature enables Java to be architecture-neutral, or to use an alternative term, platform-independent. With a Java virtual machine, you can write one program that will run on any platform.

Java's initial success stemmed from its Web-programming capability. You can run Java applets from a Web browser, but Java is for more than just writing Web applets. You can also run stand-alone Java applications directly from operating systems, using a Java interpreter. Today, software vendors usually develop multiple versions of the same product to run on different platforms. Using Java, developers need to write only one version that can run on every platform.

Because Java is architecture-neutral, Java programs are portable. They can be run on any platform, without being recompiled.

Key Words

clarity	明确，明晰
corrupt	毁坏，损坏
distributed	分布式的
dominant	占优势的，支配的
dynamic	动态的
elitist	上等的，高级的，杰出人物的
encapsulation	封装
exception	异常（事件），意外
executable	可执行的，实行的
garbage	垃圾，无用信息
inherent	内在的，固有的
mainstream	主流
modularity	模块化，模块性
multithreaded	多线程的
paradigm	范式，范例
polymorphism	多态性，多形性
prone	有……倾向的，易于……的
reusability	可复用的，可重用的
robust	健壮的，鲁棒的

Notes

[1] Since networking capability is inherently integrated into Java, writing network programs is like sending and receiving data to and from a file.

说明：本句中的"writing network programs"是主语，由"since"引导的是原因状语从句。

译文：由于联网能力一开始即被结合进Java，所以所有编写网络程序宛如向一个文件发送或从一个文件接收数据。

[2] If you run the program on other machine, it has to be recompiled on the native machine.

说明：本句中的"If you run the program on other machine"是条件状语。

译文：如果在其他机器上运行程序，则得在本机上重新编译程序。

[3] Java forces the programmer to write the code to deal with exceptions.

说明：本句中的"to write the code…"是宾语补足语。

译文：Java迫使程序员编写用于处理异常的代码。

[4] Java can catch and respond to an exceptional situation so that the program can continue its normal execution and terminate gracefully when a runtime error occurs.

说明：本句由"so that"引导目的状语，"when"引导的是时间状语。

译文：Java可捕捉异常情况并对其做出反应，以便在发生运行期错误时，程序能够继续正常运行，并从容终止。

2.4.2　Text B

Object-Oriented Programming (OOP)

The fundamental principle of object-oriented programming is that a computer program is composed of a collection of individual units, or objects which can function like sub-programs. To make the overall computation happen, each object is capable of receiving messages, processing data, and sending messages to other objects. In short, the objects can interact through their own functions (or methods) and their own data.

In this way, messages can be handled, as appropriate, by one chunk of code or by many in a seamless way. It is claimed that this gives more flexibility over simple step-by-step programming, called imperative programming or structured programming in the field of computer science.

Proponents of OOP also claim that OOP is more intuitive and that it is easier to learn, for those new to computer programming, than previous approaches. In OOP, objects are simple, self contained and easily identifiable. This modularity allows the program parts to correspond to real aspects of the problem and thereby to model the real world. Object-oriented programming often begins from a written statement of the program situation. Then by a process of inserting objects or variables for nouns, methods for verbs and attributes for adjectives, a good start is make on a framework for a program that models, and deals with that situation. This allows one to learn how to program in object-oriented languages.

Inheritance in object-oriented programming allows a class to inherit properties from a class of objects. The parent class serves as a pattern for the derived class and can be altered in several ways. If an object inherits its attributes from multiple parents, it is called multiple inheritance. Inheritance is an important concept since it allows reuse of a class definition without requiring major code changes. Inheritance encourages the reuse of code since child classes are extensions of parent classes.

Another important object-oriented concept that relates to the class hierarchy is that common messages can be sent to the parent class objects and all derived subclass objects. In formal terms, this is called polymorphism.

Polymorphism allows each subclass object to respond to the message format in a manner appropriate to its definition. Imagine a class hierarchy for gathering data. The parent class might be responsible for gathering the name, social security number, occupation, and number of years of employment for an individual. You could then use child classes to decide what additional information would be added based on occupation. In one case a supervisory position might include yearly salary, while in another case a sales position might include an hourly rate and commission information. Thus, the parent class gathers general information common to all child classes while the child classes gather additional information relating to specific job descriptions. Polymorphism allows a common data-gathering message to be sent to each class. Both the parent and child classes respond in an appropriate manner to the message.

Polymorphism gives objects the ability to responds to messages from routines when the object's exact type is not known. In C++ this ability is a result of late binding. With late binding, the addresses are determined dynamically at run time, rather than statically at compile time, as in traditional compiled languages. This static method is often called early binding. Function names are replaced with memory addresses. You accomplish late binding by using virtual functions. Virtual functions are defined in the parent class when subsequent derived classes will overload the function by redefining the function's implementation. When you use virtual functions, messages are passed as a pointer that points to the object instead of directly to the object.

Virtual functions utilize a table for address information. The table is initialized at run time by using a constructor. A constructor is invokes whenever an object of its class is created. The job of the constructor here is to link the virtual function with the table of address information. During the compile operation, the address of the virtual function is not known; rather, it is given the position in the table of addresses that will contain the address for the function.

Key Words

child class	子类
chunk	相当大的数量
computation	计算，估计

constructor	构造符，建造者
data-gathering	数据收集
imperative	命令的，必要的，势在必行的
late binding	迟绑定
occupation	职业，工作
routine	例行的，常规的，例程
seamless	无缝的
supervisory	监督的，管理的

2.4.3 Exercises

1. Translate the following phrases into English

(1)白皮书
(2)存储器分配
(3)虚函数
(4)传统的编译型语言
(5)本机代码

2. Translate the following phrases into Chinese

(1)Web browser
(2)portable application software
(3)source code
(4)runtime exception handling
(5)Java virtual machine

3. Identify the following to be True or False according to the text

(1)Virtual functions do not utilize a table for address information.
(2)A constructor is invokes whenever an object of its class is created.
(3)You accomplish late binding by using computation.
(4)Java has helped object-oriented technology enter the mainstream of computing.
(5)Distributed computing does not involve several computers working together on a network.

4. Reading Comprehension

(1)During the compile operation, the _____ of the virtual function is not known; rather, it is given the position in the table of addresses that will contain the address for the function.

    a. address
    b. name
    c. data
    d. attribute

(2) Inheritance in object-oriented programming allows a class to inherit _____ from a class of objects.

a. properties

b. data

c. approaches

d. methods

(3) One of the central issues in _____ development is how to reuse code.

a. hardware

b. software

c. system

d. structure

(4) The programs are compiled into the Java virtual machine code called _____.

a. data

b. code

c. word

d. byte-code

2.5 Database Technologies

2.5.1 Text A

A database-management system (DBMS) consists of a collection of interrelated data and a set of programs to access those data. The collection of data, usually referred to as the database, contains information about one particular enterprise. The primary goal of a DBMS is to provide an environment that is both convenient and efficient to use in retrieving and storing database information.

Database systems are designed to manage large bodies of information. The management of data involves both the definition of structures for the storage of information and the provision of mechanisms for the manipulation of information. In addition, the database system must provide for the safety of the information stored, despite system crashes or attempts at unauthorized access.[1] If data are to be shared among several users, the system must avoid possible anomalous results. The importance of information in most organizations—which determines the value of the database—has led to the development of a large body of concepts and techniques for the efficient management of data.

The storage structure and access methods used by the database system are specified by a set of definitions in a special of DDL called a data storage and definition language.[2] The result of compilation of these definitions is a set of instructions to specify the implementation details of the database schemas-details are usually hidden from the users. A database schema is also specified by DDL. The result of compilation of DDL statements is a set of tables that is stored in a special file called data dictionary, or data directory. A data dictionary is a file that contains metadata-that is, data about data. This file is consulted before actual data are read or modified in the database system.

Transaction Management

A transaction is a collection of operations that performs a single logical function in a database application. Each transaction is a unit of both atomicity and consistency. Thus, we require that transactions do not violate any database-consistency constraints. That is, if the database was consistent when a transaction started, the database must be consistent when the transaction successfully terminates. However, during the execution of a transaction, it may be necessary temporarily to allow inconsistency. This temporary inconsistency, although necessary, may lead to difficulty if a failure occurs.

Storage Management

Database typically requires a large amount of storage space. Corporate databases are usually measured in terms of gigabytes or, for the largest databases, terabytes of data. A gigabyte is 1000 megabytes or (1 billion bytes), and a terabyte is 1 million megabytes (1 trillion bytes). Since the main memory of computers cannot store this much information, the information is stored on disks. Data are moved between disk storage and main memory as needed. Since the movement of data to and from disk is slow relative to the speed of the central processing unit, it is impetrative that the database system structures the data so as to minimize the need to move data between disk and main memory.

The goal of a database system is to simplify and facilitate access to data.[3] High-level views help to achieve this goal. Users of the system should not be burdened unnecessarily with the physical details of the implementation of the system. Nevertheless, a major factor in a user's satisfaction or lack thereof with a database system is that system's performance. If the response time for a request is too long, the value of the system is diminished.[4] The performance of a system depends on what the efficiency is of the data structures used to represent the data in the database, and on how efficiently the system is able to operate on these data structures. As is the case elsewhere in computer systems, a tradeoff must be made not only between space and time, but also between the efficiency of one kind of operation and that of another.

A storage manager is a program module that provides the interface between the low-level data stored in the database and the application programs and queries submitted to the system. The storage manager is responsible for the interaction with the file manager. The raw data are stored on the disk using the file system, which is usually provided by a conventional operating system. The storage manager translates the various DML statements into low-level file-system commands. Thus, the storage manager is responsible for storing, retrieving, and updating of data in the database.

Key Words

anomalous	不规则的，反常的
burden	负担，责任
compilation	编辑，编纂
consistent	一致的，坚持的
diminish	使减少，使变小
elsewhere	在别处，到别处
impetrative	祈求，恳求
implementation	实行，履行
inconsistency	不一致性
interaction	合作，互动，互相影响
interrelate	（使）相互关联
metadata	元数据
provision	提供，收集，预备，准备
terabyte	太字节，万亿字节
tradeoff	交换，交易，折中
trillion	万亿
unauthorized	未经许可的，未授权的

Notes

[1] In addition, the database system must provide for the safety of the information stored, despite system crashes or attempts at unauthorized access.

说明：此句中的"despite"引导的是让步状语从句。

译文：另外，数据库系统还必须保证所存储信息的安全性，即使在系统崩溃或有人企图越权访问时也应保证信息的安全性。

[2] The storage structure and access methods used by the database system are specified by a set of definitions in a special of DDL called a data storage and definition language.

说明："The storage structure and access methods"是本句的主语，而"used by the database system"是主语的定语。

译文：数据库系统所使用的存储结构和访问方式通过一系列特殊的 DDL 语句来定义，这种特殊的 DDL 语句称为数据存储定义语言。

[3] The goal of a database system is to simplify and facilitate access to data.

说明："to simplify and facilitate access to data"是表语，"a database system"是定语，修饰主语"the goal"。

译文：数据库系统的目标是要简化和辅助数据访问。

[4] If the response time for a request is too long, the value of the system is diminished.

说明：本句中，"If the response time for a request is too long"是条件状语从句，"the system"是主语的定语。

译文：如果一个请求的响应速度太慢，系统的价值就会下降。

2.5.2　Text B

A data warehouse is a repository (or archive) of information gathered from multiple sources, stored under a unified schema, at a single site. Once gathered, the data are stored for a long time, permitting access to historical data. Thus, data warehouses provide the user a single consolidated interface to data, making decision-support queries easier to write. Moreover, by accessing information for decision support from a data warehouse, the decision maker ensures that on-line transaction-processing systems are not affected by the decision-support workload.

Figure 2-1 shows the architecture of a typical data warehouse, and illustrates the gathering of data, the storage of data, and the querying and data-analysis support. Among the issues to be addressed in building a warehouse are the following:

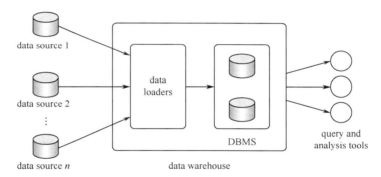

Fig. 2-1　Typical data warehouse architecture

- When and how to gather data. In a source-driven architecture for fathering data, the data sources transmit new information, either continually, as transaction processing takes place, or periodically, such as each night. In a destination-driven architecture, the data warehouse periodically sends requests for new data to the sources.

Unless updates at the sources are replicated at the warehouse via two-phase commit, the warehouse will never be quite up to date with the sources. Two-phase commit is usually far to expensive to be an option, so data warehouses typically have slightly out-of-date data. That, however, is usually not a problem for decision-support systems.

- What schema to use. Data sources that have been constructed independently are likely to have different schemas. In fact, they may even use different data models. Part of the task of a warehouse is to perform schema integration, and to convert data to the integrated schema before they are stored. As a result, the data stored in the warehouse are not just a copy of the data at the sources. Instead, they can be thought of as a stored view (or materialized view) of the data at the sources.

- How to propagate updates. Updates on relations at the data sources must be propagated to the data warehouse. If the relations at the data warehouse are exactly the same as those at the data source, the propagation is straightforward.
- What data to summarize. The raw data generated by a transaction-processing system may be too large to store on-line. However, we can answer many queries by maintaining just summary data obtained by aggregation on a relation, rather than maintaining the entire relation. For example, instead of storing data about every sale of clothing, we can store total sales of clothing by category.

Key Words

aggregation	聚集，集成，集结
category	种类，类别，类型
commit	提交，承诺
propagate	扩散，繁荣，繁衍
query	查询，质疑
repository	仓库，仓储室
straightforward	直接的，坦率的，明确的
transaction	事务，处理，交易
warehouse	仓库，货栈，批发商店
workload	工作量，负载

2.5.3 Exercises

1. Translate the following phrases into English

(1)数据库管理系统

(2)数据定义语言

(3)数据字典

(4)物理细节

(5)完整性约束

2. Translate the following phrases into Chinese

(1)database system

(2)database schema

(3)storage manager

(4)data warehouse

(5)decision-support

3. Identify the following to be True or False according to the text

(1)Databases are usually measured in terms of MB or GB.

(2)The goal of a database system is not to simplify access to database.

(3) A storage manager provides the interface between a database and a application program.

(4) A data definition language can be used to define a database schema.

(5) The data values stored in the database must satisfy certain consistency constraints.

4. Reading Comprehension

(1) Data warehouses provide _____ a single consolidated interface to data, making decision-support queries easier to write.

a. the storage structure

b. DBA

c. computer

d. the user

(2) The raw data are stored on the disk using the _____.

a. data dictionary

b. file system

c. DBMS

d. DBA

(3) Each _____ is a unit of both atomicity and consistency.

a. transaction

b. database

c. storage structure

d. schema details

(4) _____ is designed to manage large bodies of information.

a. a file system

b. a transaction

c. a database system

d. a database language

2.6 Software Engineering

2.6.1 Text A

Early approaches to software engineering insisted on performing analysis, design, implementation, and testing in a strictly sequential manner. As a result, software engineers insisted that the entire analysis of the system be completed before beginning the design and, likewise, that the design be completed before beginning implementation. The result was a development process now referred to as the waterfall model, an analogy to the fact that the development process was allowed to flow in only one direction.

The first phase, requirements definition, refers to the period during which the requirements of the system desired, that is, it's functional characteristics and operational details, are specified.[1] If the system is to be a generic product sold in a competitive market, this analysis would involve a broad-based investigation to identify the needs of potential customers. If, however, the system is to be designed for a specific user, then the process would be a more narrow investigation.[2] As the needs of the potential user are identified, they are compiled to form a set of requirements that the new system must satisfy. These requirements are stated in terms of the application rather than in the technical terminology.

One requirement might be that access to data must be restricted to authorized personnel. Another might be that the data must reflect the current state of the inventory as of the end of the last business day or that the arrangement of the data as displayed on the computer screen must adhere to the format of the paper forms currently in use.[3] After the system requirements are identified, they are converted into more system technical specifications. An error in requirements, for example, a misstated function, leads to a faulty design and an implementation that does not do what is required. If this is allowed to proceed undetected, say, until the testing phase, the cost of repairing this error (including redesign and re-implementation) can be substantial.

The second phase is design. Design concentrates on how the system will accomplish the goals. It is here that the structure of the software system is established. The input to this phase is a (debugged and validated) requirements document: the output is a design expressed in some appropriate form (for example, pseudo-code). Validation of a design is important. Each requirement in the requirements document must have a corresponding design fragment to meet it. Formal verification, while possible to a limited extent, can be exceedingly difficult. More informal revolve the entire design team, management, and even the client.

It is a well-established principle that the best structure for a large software system is a modular one. Indeed, it is by means of this modular decomposition that the implementation of large systems becomes a possibility. Without such a breakdown, the technical details required in the implementation of a large system would exceed a human's comprehensive powers. With a modular design, however, only the details pertaining to the module under consideration need be mastered. This same modular design is also conductive to future maintenance because it allows changes to be made on a modular basis.

The third phase, implementation, is the actual coding of the design developed in the second phase. The design must be translated into a machine-readable form. The coding step performs this task. If design is performed in a detailed manner, coding can be accomplished mechanistically. The lure of this phase is strong, and many a foolhardy programmer has been drawn to it before adequately laying the groundwork in the first two phases. As a result, requirements are incompletely understood and the design is flawed. The implementation proceeds blindly, and many problems arise as a result.

The forth phase is software testing. Testing is closely associated with implementation, because each module of the system is normally tested as it is implemented. In the development of a large system, testing involves several stages. First, each program module is tested as a single program, usually isolated from the other programs in the system. Such testing is defined as module testing or unit testing. Unit testing is done in a controlled environment whenever possible so that the test team can feed a predetermined set of data to the module being tested and observe what output data are produced.[4] In addition, the test team checks the internal data structures, the logic, and the boundary conditions for the input and output data.

When collections of modules have been unit-tested, the next step is to insure that the interfaces among the modules are defined and handled properly. Integration testing is the process of verifying that the components of a system work together as described in the program design and system design specifications. Once we are sure that information is passed among modules according to the design prescriptions, we test the system to assure that it has the desired functionality. A function test evaluates the system to determine if the functions described by the requirements specification are actually performed by the integrated system.

Key Words

analogy	相似，类似，比拟
breakdown	崩溃，损坏
competitive	竞争的，对立的，竞赛的
decomposition	分解，解体
exceedingly	非常，极其，极度地
faulty	不完善的，有错误的
foolhardy	蛮干的
fragment	分段，碎片，片段
groundwork	基础，基本原理，地基
inventory	清查，存货清单，编制……的目录
investigation	调查，研究
likewise	同样，也，而且
mechanistically	机械地
prescription	指示，训令，法规，规则
pseudo	假的，伪的，冒充的
readable	易读的，易懂的
restrict	限制，约束
sequential	按次序的，顺序的，序列的
substantial	充实的，实质的
terminology	专有名词，术语
verification	证明，证实，核实

Notes

[1] The first phase, requirements definition, refers to the period during which the

requirements of the system desired, that is, it's functional characteristics and operational details, are specified.

说明:"requirements definition"作为主语"The first phase"的同位语,"that is, it's functional…"是定语。

译文:第一阶段是需求定义阶段,是指系统期望的需求分析,即描述功能特征和操作细节的阶段。

[2] If, however, the system is to be designed for a specific user, then the process would be a more narrow investigation.

说明:本句由"If"引导条件状语从句。

译文:但是,如果系统是为特殊用户设计的,那么这个过程就是一个更专业的调查。

[3] Another might be that the data must reflect the current state of the inventory as of the end of the last business day or that the arrangement of the data as displayed on the computer screen must adhere to the format of the paper forms currently in use.

说明:本句有两个并列的表语从句"that the data must…"和"that the arrangement…",而"as of the end of the last business day"是时间状语。

译文:另一种需求可能是当最后一个工作日结束时,数据必须反映目前的清单状态,或者可能是在计算机屏幕上的数据组织必须按照目前使用的纸质形式的格式来显示。

[4] Unit testing is done in a controlled environment whenever possible so that the test team can feed a predetermined set of data to the module being tested and observe what output data are produced.

说明:本句由"so that"引导目的状语从句。

译文:在任何可能的时候,单元测试是在控制环境下进行的,这是为了让测试小组能提供被测试模块先前确定的数据并观察产生的输出数据。

2.6.2 Text B

No matter how capably we write programs, it is clear from the variety of possible errors that we should check to insure that our modules are coded correctly. Many programmers view testing as a demonstration that their programs perform properly. However, the idea of demonstrating correctness is really the reverse of what testing is all about. We test a program in order to demonstrate the existence of an error. Because our goal is to discover errors, we can consider a test successful only when an error is discovered. Once an error is found,"debugging"or error correction is the process of determining what causes the error and of making changes to the system so that the error no longer exists.

Any experienced programmer mentally tests each line as it is produced and mentally simulates the execution of any module prior to any formal testing stage. A "successful" test run means only that no errors were uncovered with the particular circumstances tested; it says nothing about other circumstances. In theory, the only way that testing can show that a program is correct is if all possible cases are tried (known as an exhaustive test), a situation technically impossible for even the simplest programs. Suppose, for example, that we have written a program to compute the average grade on an examination. An exhaustive test would require all possible combinations of marks and class sizes; it could take many years to complete the test.

Recall that the requirements were specified in two ways: first in the customer's terminology and again as a set of software and hardware requirements. The function test compares the system being built with the functions described in the software and hardware requirements. Then, a performance test compares the system with the remainder of the software and hardware requirements. If the test is performed in the customer's actual working environment, a successful test yields a validated system. However, if the test must be performed in a simulated environment, the resulting system is a verified system.

When the performance test is complete, we as developers are certain that the system functions according to our understanding of the system description. The next step is to confer with the customer to make certain that the system works according to the customer's expectations. We join with the customer to perform an acceptance test in which the system is checked against the customer's requirements description. When the acceptance test is complete, the accepted system is installed in the environment in which it will be used; a final installation test is performed to make sure that the system still functions as it should.

Maintenance begins after the system is released. Maintenance activities involve making enhancements to software products, adapting products to new environments, and correcting problems. Software product enhancement may involve providing new functional capabilities, improving user displays and modes of interaction, upgrading external documents and internal documentation, or upgrading the performance characteristics of a system, Adaptation of software to a new environment may involve moving the software to a different machine, or for instance, modifying the software to accommodate a new telecommunications protocol or an additional disk drive. Problem correction involves modification and revalidation of software to correct errors. Some errors require immediate attention, some can be corrected on a scheduled, periodic basis, and others are known but never corrected.

Key Words

acceptance	接受，同意
boundary	边界，界限
confer	比较，对照，商议，磋商
feed	满足，提供，注入
isolate	隔离，孤立
recall	回忆，回顾
reverse	相反，背面
validate	证实，确认

2.6.3 Exercises

1. Translate the following phrases into English

(1)内部数据结构

(2)仿真环境

(3)维护阶段

(4)软件开发

(5)集成测试

2. Translate the following phrases into Chinese

(1)software engineering

(2)waterfall model

(3)software testing

(4)boundary condition

(5)requirements specification

3. Identify the following to be True or False according to the text

(1)We test a program in order to demonstrate the existence of an error.

(2)The function test compares the system being built with the functions described in the software and hardware requirements.

(3)Design doesn't concentrate on how the system will accomplish the goals.

(4)When the acceptance test is complete, the accepted system is installed in the environment in which it will be used.

(5)Before the system requirements are identified, they are converted into more system technical specifications.

4. Reading Comprehension

(1)_____is the process of verifying that the components of a system work together as described in the program design and system design specifications.

a. Function testing

b. Unit testing

c. Integration testing

d. Module testing

(2)_____evaluates the system to determine if the functions described by the requirements specification are actually performed by the integrated system.

a. A function test

b. A unit test

c. A module test

d. A task test

(3)If the test is performed in the customer's_____, a successful test yields a validated system.

    a. laboratory environment

    b. virtual environment

    c. simulated environment

    d. actual working environment

(4)Maintenance activities involve making enhancements to software products, adapting products to new environments, and _____.

    a. find next error

    b. correcting problems

    c. find errors

    d. select problems

2.7 计算机专业英语的阅读与翻译

2.7.1 阅读方法

阅读实际上就是语言知识、语言技能和智力的综合运用。在阅读过程中，这三个方面的作用总是浑然一体，相辅相成的。扎实的语言基础是提高阅读能力的先决条件。首先，词汇是语言的建筑材料。要想提高专业英语资料的阅读能力则必须扩大词汇量，尤其是掌握一定量的计算机专业词汇。若词汇量掌握得不够，阅读时就会感到生词多，不但会影响阅读的速度，而且会影响理解的程度，从而不能进行有效的阅读。其次，语法是语言中的结构关系，按一定的规则把词或短语组织到句子中，以表达一定的思想。熟练掌握英语语法和惯用法也是阅读理解的基础。

阅读时，注意每次视线的停顿应以一个意群为单位，而不应以一个单词为单位。若逐个单词地读，当读完一个句子或一个段落时，早就忘记前面读的是什么内容了。这样读不仅速度慢，还会影响理解。因此，正确的阅读方法可以提高阅读速度并提高阅读理解能力。

常用的有效的阅读方法有三种，即略读（skimming）、查读（scanning）和精读（reading for full understanding）。

略读指以尽可能快的速度进行阅读，了解文章的主旨和大意，对文章的结构和内容获得总的概念和印象。在进行略读时，必须特别集中精力，注意短文的开头句和结尾句，力求抓住文章的主旨大意；注意文章的体裁和写作特点，了解文章结构；注意支持主题句或中心思想的信息句。

查读的目的主要是有目的地去找出文章中某些特定的信息，也就是说，在对文章有所了解的基础上，在文章中查找与某一问题、某一观点或某一单词有关的信息。查读时，要以很快的速度扫视文章，确定所查询的信息范围，注意所查信息的特点，如有关日期、专业词汇、某个事件、某个数字、某种观点等，寻找与此相关的关键词或关键段落。注意：与所查信息无关的内容可以略过。

精读指仔细地阅读，力求对文章有深层次的理解，以获得具体的信息，包括理解衬托主题句的细节，根据作者的意图和中心思想进行推论，根据上下文猜测词义等。对难句、长句，要借助语法知识进行分析，准确地理解。

总之，要想提高阅读理解能力，必须掌握以下6项基本的阅读技能。
- ◆ 掌握所读材料的主旨大意。
- ◆ 了解阐述主旨的事实和细节。
- ◆ 根据上下文判断某些词汇和短语的含义。
- ◆ 既理解个别句子的意义，也理解上下文之间的逻辑关系。
- ◆ 根据所读材料进行一定的判断、推理和引申。
- ◆ 领会作者的观点、意图和态度。

科技文章的内容是浓缩的，在任何一页上都会发现大量的信息。可以把一篇文章分成几个小部分、几页来读，甚至一次就读一个自然段。在阅读计算机科技文章时，还要注意在阅读中及时获取细节信息。在所有的文章中，作者都使用细节或事实来表达和支持他们的观点。有时还要区分哪些是重要细节，哪些是次要细节。

阅读能力的提高离不开阅读实践。在打好语言基本功的基础上，还要进行大量的阅读实践。词汇量和阅读能力的提高是一种辩证关系：要想读得懂，读得快，就必须扩大词汇量；反之，要想扩大词汇量，就必须大量阅读。同样，语法和阅读之间的关系也是如此：牢固的语法知识能够促进阅读的顺利进行，提高阅读的速度和准确率；反之，通过大量的阅读实践又能够巩固已掌握的语法知识。只有在大量的阅读中，才能培养语感，掌握正确的阅读方法，提高阅读理解能力。同时在大量的阅读中，还能巩固计算机专业知识及了解到计算机专业的发展趋势，对于跟踪计算机技术的发展很有好处。

在阅读专业文章时，可以借鉴以下几种阅读技巧。
- ◆ 鉴别阅读法

鉴别阅读法是一种快速提炼文章的段意、主要内容和中心思想的阅读方法。鉴别阅读法包括三个环节：划分出文章的段落，迅速找出段落的中心句、重点句，或用自己的语言概括出段意；连接各段的段意，分析文章的重点句、段，归纳主要内容；在阅读过程中，要留意文章的题目、开头段、结尾段及文章的议论部分，从中概括出中心思想。

鉴别阅读法实际上是通过以上三个环节来掌握文章的重要信息，因而无须一字不漏地通读全文。在运用鉴别阅读法时，注意力要高度集中，使大脑处于积极思维状态，这样才能保证阅读质量。

- ◆ 默读法

在阅读时，大脑直接感受文字的意思，不必通过发音器官将文字转换为声音，这种阅读方式就叫无声阅读。在进行无声阅读时，由于发音器官受抑制，视觉不受逐字换音的牵制，因而视角广度大，便于以句、以行甚至以段为单位进行阅读，还可以根据阅读目的的需要进行浏览、跳读。由于是直接理解文字的意义，省略了发音阶段，所以无声阅读的速度比出声阅读的速度快得多。

- ◆ 视觉辅助阅读法

很多阅读者常常在阅读时用手指点着单词，传统上把这种习惯看成一种错误，并要求他们把手指从书本上拿开。现在看来，要做的不是让他们把手指从书本上拿开，而是让他们更快地移动手指。显然，手指不会减缓眼睛的移动，相反，它在帮助读者养成流畅的阅读节奏的习惯方面有着不可估量的作用。当然，读者也不必局限于使用手指作为视觉引导物，也可以用钢笔或铅笔引导视觉。对于还没有掌握快速阅读技巧的阅读者来说，可以充分利用视觉辅助引导物进行快速阅读训练。

2.7.2 翻译的方法与过程

要想做好翻译工作，必须从深刻理解原文入手，力求做到确切表达译文。原文是翻译的出发点和唯一依据，只有彻底理解原文的含义，才有可能完成确切的翻译，才能达到上述翻译标准的要求。要想深刻理解原文，首先应认识到专业科技文献所特有的逻辑性、正确性、精密性和专业性等特点，力求从原文所包含的专业技术内容方面去加以理解。其次，要根据原文的句子结构，弄清每句话的语法关系，采用分组归类的方法辨明主语、谓语、宾语及各种修饰语，联系上下文来分析和理解句与句之间、主句与从句之间的关系。

1. 专业词汇的翻译

如果把一种语言的所有词汇作为一个词汇总集来看待，则各种词汇的分布情况和运用频率是不一样的。在词汇的总体分布中，有些词属于语言的共核部分，如功能词和日常用词。这些词构成了语言的基础词汇。此外，各个学术领域的技术术语和行业词构成了词汇总集的外缘，而处于基础词汇和外缘之间的是那些准技术词汇（sub-technical words）。各个学科领域还存在大量的行业专用表达方法和词汇，正是这些词汇在双语翻译中形成了真正的难度。在科技翻译中，准确性是第一要素；如果为追求译文的流畅而牺牲准确性，不但会造成科技信息的丢失，影响文献交流，还可能引起误解，造成严重后果。

常用翻译词汇的方法如下。

◆ 音译

音译指根据英语单词的发音译成读音与原词大致相同的汉字。通常，表示计量单位的词和一些新发明的材料或产品的汉语名称在刚开始时基本就是音译的。或在出于某些原因不便采用意译法时，可采用音译法或部分音译法。

例：Radar 是取 radio detection and ranging 等词的部分字母拼成的，若译成"无线电探测距离设备"，则显得十分啰唆，故采用音译法，译成"雷达"。

又如：

baud	波特（发报速率单位）
bit	比特（二进制信息单位）
hertz	赫兹（频率单位）
penicillin	盘尼西林（青霉素的旧称）
vaseline	凡士林

◆ 意译

意译指对原词所表达的具体事物和概念进行仔细推敲，以准确译出该词的科学概念。这种译法最为普遍，科技术语在可能情况下应尽量采用意译法。采用这种方法便于读者顾名思义，不加说明就能直接理解新术语的确切含义。

例：
loudspeaker	扬声器
semiconductor	半导体
videophone	可视电话
E-mail = Electronic mail	电子邮件
modem = modulator + demodulator	调制解调器

◆ 形译

以英语常用字母的形象来为形状相似的物体定名，翻译时也可以通过具体形象来表达原义，也称为"象译"。科技文献常涉及型号、牌号、商标名称及代表某种概念的字母。这些一般不必译出，直接抄下即可。另外，对于人名及公司名等名称类的词汇，翻译时可直接使用原文。

例：
I-shaped	工字形
T square	丁字尺
C network	C 形网络
X ray	X 射线
Y-connection	Y 形连接
Zigzag wave	锯齿形波

◆ 词义的选择与引申

在翻译过程中，若英、汉语词汇都是相互对应的单义词时则汉译不成问题，如 ferroalloy（铁合金）。然而，由于英语词汇来源复杂，一词多义和一词多性的现象十分普遍，如 power 在数学中译为"乘方"，在光学中译为"率"，在力学中译为"能力"，在电学中译为"电力"。

例：The electronic microscope possesses very high resolving power compared with the optical microscope.

译文：与光学显微镜相比，电子显微镜具有极高的分辨率。

例：Energy is the power to do work.

译文：能量是指做功的能力。

英、汉两种语言在表达方式、方法上的差异较大，英语一词多义现象使得在汉语中很难找到绝对相同的词。如果仅按词典意义原样照搬，不仅使译文生硬晦涩，而且可能会词不达意，造成误解。因此，有必要结合语言环境透过外延看内涵，对词义做一定程度的扩展、引申。

例：Two and three make five.

译文：二加三等于五。（make 本意为"制造"，这里扩展为"等于"。）

The report is happily phrased.

译文：报告措辞很恰当。（happily 不应译为"幸运地"。）

◆ 词语的增减与变序

由于英语和汉语各自独立演变发展，因而在表达方法和语法结构上有很大的差别。在英译汉时，不可能要求二者在词的数量上绝对相等。通常，应依据句子的意义和结构适当地增加、减少或重复一些词,以使译文符合汉语习惯。

例:The more energy we want to send, the higher we have to make the voltage.

译文:要想输电更多,电压也就得更高。(省略 we。)

例:This condenser is of higher capacity than is actually needed.

译文:这只电容器的容量比实际所需要的容量大。(补译省略部分的 capacity。)

英语和汉语的句子顺序通常都是按主语+谓语+宾语排列的,但修饰语的区别却较大。英语中各种短语或定语从句作为修饰语时,一般都是后置的,而汉语的修饰语几乎都是前置的,因而在翻译时应改变动词的顺序。同时,还应注意英语几个前置修饰语(通常为形容词、名词和代词)中最靠近被修饰词的为最主要的修饰语,翻译时应首先译出。此外,英语中的提问和强调也大都用倒装词序,翻译时应注意还原。

例:The transformer is a device of very great practical important which makes use of the principle of mutual induction.

译文:变压器是一种利用互感原理的在实践中很重要的装置。

◆ 语法成分的转换

为了使译文达到"明确""通顺""简练"的要求,有时需要把原语句中的某种成分转译为另一种成分。成分的转换在大多数情况下并不引起词性的转换,而词性的转换经常会引起成分的转换。

(1)主语的转译

为了使译文简明通顺,要把英语的被动句译成汉语的主动句,这时常常把英语句中的主语转译为汉语句中的宾语。有时根据译文修辞上的需要,将英语主动句的主语转译为汉语句中的宾语。这种转译中的词性一般不变。

例:Much progress has been made in computers in recent years.

译文:近年来,计算机取得了很大的进步。

例:Considering the processor range only, several disadvantages attend the computer.

译文:仅就处理器范围而言,该计算机就有几个缺点。

汉语句中的主语有时可转换成英语句中的宾语、介词宾语、表语、谓语、状语或定语。

例:Light beams can carry more information than radio signals because light has a much higher frequency than radio waves.

译文:光束运载的信息比无线电信号运载的信息多,因为光波的频率比无线电波的高。

例:It is a basic rule of design that each program is adaptable to users.

译文:设计的基本准则是每项程序都适用于用户。

例:The electronic computer is chiefly characterized by its accurate and rapid computations.

译文:电子计算机的主要特点是计算准确和迅速。

(2)谓语的转译

为了使译文符合汉语习惯,把英语句中的谓语动词转译为汉语名词并将它作为主语。有时为了使译文更简明通顺,也将英语句中的谓语转译成汉语句的宾语、定语、状语等成分。

例:This paper aims at discussing new developments in computers.

译文:本文的目的在于讨论计算机的新发展。

例：The computer schedules the operations of the whole plant.

译文：计算机能排定整套设备的操作时间表。

汉语句中的谓语有时可转译成英语句中的定语、状语、补语、宾语、主语或表语。

例：Electronic computers have great importance in the production of modern industry.

译文：电子计算机在现代工业生产中很重要。

例：These new memories are now in wide use.

译文：这些新型存储器正在被广泛使用。

（3）宾语的转译

当英语句中的动词宾语或介词宾语在意义上与主语有密切联系时，可以将这种英语句中的宾语转译成汉语句中的主语。另外，当英语名词转译为汉语动词或形容词时，可能发生宾语转译成谓语的情况。

例：You should not confuse the processor's instruction set with the instructions found in high-level programming languages, such as BASIC or PASCAL.

译文：处理器的指令集与 BASIC 或 PASCAL 这样的高级程序设计语言中的指令不应混淆。

汉语中常常使用无主句，或在一定的上下文中将主语省略。将这样的汉语句译成英语句时，句中的宾语就变成英语被动语态句式的主语。有时，汉语句中的宾语还可以转译成英语句中的状语、定语或表语。

例：Microcomputers have found their application in the production of genius sensors.

译文：微型计算机已经应用到智能传感器的生产中。

例：To fulfill this need, many kinds of printers have been designed.

译文：为了满足这种需求，已经设计出许多类型的打印机。

例：Dot-matrix printers are highly reliable and inexpensive.

译文：点阵打印机具有可靠性强和价格便宜的特点。

（4）定语的转译

英语句中常常利用定语来表达事物的性质、特点和参数，而汉语句中却无此习惯。因此，遇到这种情况时，可将英语句中的定语转译成汉语句中的表语。另外，在将充当名词定语的形容词转译为动词时，经常发生将定语转译为谓语的情况。也可将充当定语的分词、分词短语或介词短语转译为谓语。

例：The fast speed is one of the advantages of this printer.

译文：这台打印机的优点之一是速度快。

例：There is a large amount of paper wasted due to the hard-copy output.

译文：硬拷贝输出浪费大量的纸张。

汉语句中的定语一般放在它所修饰的词语之前，而放在英语句中则可前可后。有时，汉语句中的定语较长，如不变为英语句中的后置定语，还可译为状语。此外，汉语句中的定语有时还可以转译为英语句中的谓语、宾语或主语。

例：In data-processing operation, the final results must be made available in a form usable by humans.

译文：数据处理操作的最终结果必须成为可供人们使用的形式。
例：Matrix printers are relative low cost, high speed, and quite operation.
译文：点阵式打印机的价格较低，速度较快，噪声较小。

（5）状语的转译

当将英语句中的副词转译为汉语句中的名词时，经常发生将状语转译为主语的情况。充当状语修饰谓语动词的介词短语，有时也可以转译为汉语句中的主语。另外，有时根据修辞的需要，把英语句中用作状语的副词转译为汉语译句的补语、定语、谓语、表语或宾语等。

例：The speeds of ink-jet printers are extremely high.
译文：喷墨打印机的速度高得惊人。
例：Computerized systems using electronic transducers can actuate automatic exposure and focusing mechanisms in cameras.
译文：使用电子传感器的计算机化系统可以启动摄像机中的自动曝光和聚焦装置。

汉语句中的状语有时可转译为英语句中的表语、定语、谓语或主语。

例：This type of display screen can provide a clearer screen image.
译文：这种显示器能够更清晰地显示屏幕图像。

2. 翻译方法与过程

科技文献本身体裁多样，而且随着时代的推进，越来越多的科技文献开始显示出其内容及文体风格上的创新和活跃。如果没有一定的翻译技巧，则翻译不出具有不同特点的科技文章。一般来说，科技文献主要述说事理、描写现象、推导公式及论证规律，因此其特点是结构严谨、逻辑严密、行文规范、用词准确、技术术语正确、修辞手段较少。

专业科技文献中的长句、难句较多，各种短语和从句相互搭配、相互修饰，使人感到头绪纷繁、无所适从。在这种情况下更应重视语法分析，突出句子骨架，采用分解归类、化繁为简、逐层推进理解的策略。

例：Just click to open the new Office E-mail header in Word and send your document as an E-mail message that retains your original formatting.
译文：只要单击就可打开 Word 中的新 Office E-mail 标题，这样就可以把你的文档资料按照原来的格式作为电子邮件发送出去。

本句使用了科技英语中常用的"祈使句+and…"句型。

例：Unlike Word for Windows, in which macros are directly linked to document and template files, AmiPro macros are contained in a separate file.
译文：在 Windows Word 环境中，宏是直接连接到文档和模板文件上的。与此不同的是，AmiPro 宏病毒则包含在一个独立的文件中。

句中的"in which macros are directly linked to document and template files"是一个介词前置的非限定性定语从句，修饰"Word for Windows"。

在翻译的过程中，理解是第一位的，表达是第二位的。

◆ 理解原文

透彻理解原文是确切表达的前提。理解原文必须从整体出发，不能孤立地看待一词一句。每种语言几乎都存在着一词多义的现象。因此，同样一个词或词组，在不同的上下文搭配中，在不同的句法结构中就可能有不同的意义。

（1）领略全文大意，分析语法关系

任何一篇文章都是一个有机整体，在翻译的过程中切忌看一句译一句，一定要在通读全文的基础上，领略原文的大意。透过各种语言现象对句子进行微观分析，即弄清句中各词、各成分之间的种种关系，如主从关系、主谓关系、动宾关系等。

（2）理解原作事理，注意逻辑判断

有些句子在语法上可以有几种不同的解释，在语义上也可以有几种不同的理解，要判断这些句子的真正含义，还必须通过逻辑判断，对句子结构、语言环境、上下文和事理关系进行综合分析才能奏效。在理解原文的事理和逻辑关系时，必须仔细推敲，反复斟酌原文的含义，有时还要估计实际情况，根据自己的生活体验和客观道理来验证自己的理解是否合乎情理。

例：The author proposes an approach to the creation of an integrated method of investigation and designing objects, based on a local computer system.

译文：作者创立一种综合法，以本地计算机系统为基础来进行对象研讨与设计。

例：The binary system of representing numbers will now be explained by making reference to the familiar decimal system.

译文：下面参照熟悉的十进制系统来说明表示数目的二进制系统。

例：The paper addresses an important problem in data-base management systems.

译文：本文讨论了基本数据管理系统的一个重要问题。

◆ 汉语表达

表达是指用适当的译文把已经理解的原文再现出来。一般来说，表达的好坏取决于译者对原文的理解程度、译者的逻辑思维能力和运用汉语的熟练程度。为了使译文准确通顺，在表达阶段一般要注意以下几个方面。

（1）表达的规范性

科技语体讲究论证的逻辑性，要求语言规范。对于科技文章来说，更要求运用规范的汉语来表达。

（2）表达的逻辑性

任何一位作者在著书立说时，都要运用概念、判断和推理这一思维形式，而人的思维要反映客观规律，则必须符合逻辑，因此，表达思维的语言也要符合逻辑。

（3）表达的主动权

英、汉两种语言在句法结构、表意方式等方面存在很大的差异。因此，为了得到准确而流畅的译文，译者就不能把自己局限在原文的语言形式中。出于汉语表达的需要，译者可以甚至必须跳出原文的框框，对原文的句子成分、结构形式进行必要的调整，从容自若地按照汉语的特点和习惯组织自己的译文。

下面这段文章讲述的是读书对人们生活的影响，翻译时应注意内容的连贯和准确。

例：Reading broadens our experience. It enables us to feel how others felt about life, even if they lived thousands of miles away and centuries ago. Although we may be unworthy, we can become the friends of wise men. Only books can give us these pleasures. Those who cannot enjoy them are poor men; those who enjoy them most obtain the most happiness from them.

译文：读书能丰富我们的阅历。读书使我们能体验到别人对生活的感受，哪怕他们生活在千里之外或数百年之前。尽管我们可能是微不足道的，但我们却能和聪明人交朋友。只有书籍能给我们这些乐趣。那些不能享受读书之乐的人是贫乏的人，而那些最喜欢读书的人，可以从中得到最大的幸福与满足。

◆ 校对阶段

校对阶段是理解和表达的进一步深化，是使译文符合标准的一个必不可少阶段，是对原文内容的进一步核实、对译文的进一步推敲。校对对于科技文章的译文来说尤为重要，因为科技文章要求高度精确，公式、数据较多，故稍微疏忽就会给工作造成严重的损失。

例：The technical possibilities could well exist, therefore, of nation-wide integrated transmission network of high capacity, controlled by computers, interconnected globally by satellite and submarine cable, providing speedy and reliable communications throughout the world.

这句话看起来令人难以理解，但若采用分解归类的语法分析方法则不难理解。首先，能够充当句子谓语的只能是 could well exist，而不可能是其他非限定动词，如 integrated、controlled、interconnected。然后，主语自然是 possibilities，由于 exist 是不及物动词，因此不存在宾语。进一步分析将发现用作定语的介词短语 of nation-wide... cable 除修饰 possibilities 外没有其他名词可以承受，而分词短语 controlled...和 interconnected...又进一步修饰介词短语中的 network。providing... the world 则显然是表示结果的状态。这样一来，句子的骨架就比较清楚了。本句所以将定语和待修饰的词分开，是因为定语太长而谓语较短，将谓语提前有助于整个句子结构的平衡。

译文：因此，在技术上完全有可能实现全国性集成传输网，这种网络的容量大，可由计算机控制，并能通过卫星和海底电缆实现全球互联，提供高速、可靠的全球通信。

在翻译科技资料时，首先要把原文全部阅读一遍，了解其内容大意、专业范围和体裁风格，然后开始翻译。如果条件许可，在动手翻译之前最好先熟悉一下有关的专业知识。遇到生词，不要马上查字典，应该先判断是属于普通用语，还是属于专业用语。如果是专业词汇，则要先分析是属于哪一个具体学科范围的，然后再查找普通词典或有关的专业词典。翻译时，最好不要看一句译一句，更不能看一个词译一个词，而应该看一小段译一小段。这样做便于从上下文联系中辨别词义，也便于注意句与句之间的衔接、段与段之间的联系，使译文通顺流畅，而不致成为一句句孤立译文的堆砌。翻译科技文献并不要求像翻译文艺作品那样在语言形象、修辞手段上下功夫，但是要求译文必须概念清楚、逻辑正确、数据无误、文字简练、语句流畅。

下面是一篇关于计算机犯罪的文章，它共有两段。第一段讲述计算机具有优良的性能，即在很大程度上取代了文书工作；并以银行系统为例讲述它的优点，即它不会有个人情绪、不会偷钱；但用一句"它也没有良知"引出有人利用计算机犯罪的可能性。第二段用一个实例讲述利用计算机犯罪的隐蔽性和严重性：一个银行职员通过转账的方法，盗窃客户的钱，数量很大，一直没有被发现，直到因为赌博案发，他利用计算机盗窃的事才被发现。

例：

In many businesses, computers have largely replaced paperwork, because they are fast, flexible, and do not make mistakes. As one banker said, "Unlike humans, computers never have a bad day." And they are honest. Many banks advertise that their transactions are "untouched by human hands" and therefore safe from human temptation. Obviously, computers have no reason to steal money. But they also have no conscience, and the growing number of computer crimes shows they can be used to steal.

Computer criminals don't use guns. And even if they are caught, it is hard to punish them because there are witness and often no evidence. A computer cannot remember who used it, it simply does what it is told. The head teller at a New York City bank used a computer to steal more than one and a half billion dollars in just four years. No one noticed this theft because he moved the money from one account to another. Each time a customer he had robbed questioned the balance in his account, the teller claimed a computer error, then replaced the missing money from someone else's account. This man was caught only because he was a gambler. When the police broke up an illegal gambling operation, his name was in the records.

译文：

在许多商业活动中，计算机在很大程度上取代了文书工作。因为它们速度快、灵活，而且不会犯错误。正如一位银行家所说："与人不同的是，计算机没有情绪不好的时候。"计算机很诚实，许多银行都在广告中说其业务往来都不是"由人手办理"的，所以不会受到个人情绪的影响。计算机没有理由去偷钱，这是显而易见的。但它们也没有良知，数量逐渐上升的计算机犯罪表明它们可以被用于盗窃。

计算机犯罪不用枪，即使被抓住了，也很少受到惩罚，因为没有证人，而且通常也没有证据。计算机无法记住谁使用过它们，它们只是执行命令。纽约市银行的一个出纳主任在短短四年时间内使用计算机偷了15亿多美元。没人注意到这个盗窃案，因为他把钱从一个账户转到另一个账户。每次，当被他偷过的顾客对自己账户的结算余额提出疑问时，这个出纳员就说这只是计算机的小错误，然后他把另一个账户的钱补到这个账户上，后来，这个人被逮捕了，只是因为他赌博，当警方捣毁一个非法赌博机构时，他的名字被记录在案。

下面短文讲述的是计算机软件，涉及计算机软件的定义及功能。在对一些词的理解上就要注意准确性及合理性。例如，inanimate device、intelligence、instructions、software、programs、software package 及 documentation 可分别翻译成无生命的装置、智能、指令、软件、程序、软件包及文档。在对这篇短文的关键词理解的基础上，我们可以给出每句话的表达。到此，可能还存在对原文表达不准确的地方，因此需要对原文内容做进一步的核实和推敲。

例：A computer is an inanimate device that has no intelligence of its own and must be supplied with instructions so that it knows what to do and how and when to do it. These instructions are called software. The importance of software can't be overestimated. Software is made up of a group of related programs, each of which is a group of related instructions that perform very specific processing tasks. Software acquired to perform a general business function is often referred to as a software package. Software packages, which are usually created by professional software writers, are accompanied by documentation—that explains how to use the software.

译文：一台计算机本是一个无生命的装置。它没有自身的智能，并且必须给它指令，它才能知道去做什么、如何做及何时去做。这些指令称为软件。对软件的重要性无论如何估计，都不会过分。软件是由一组相互有关的程序组成的，其中的每个程序都是一组相关指令，而每组指令都将执行极为特定的处理任务。用来完成一般事务性功能任务的软件通常称为软件包。软件包一般均由软件专业人员编写。软件包都带有文档，文档用来解释如何使用这个软件。

2.8 习题答案与参考译文

2.8.1 第 2 单元习题答案

2.1 Data Structure

1. Translate the following phrases into English

(1)内部结构		internal construction
(2)数据类型		data type
(3)结构数据类型		structured data type
(4)算术表达式		arithmetic expression
(5)函数调用		function call

2. Translate the following phrases into Chinese

(1) last in/first out　　　　　　后进先出
(2) atomic data type　　　　　　原子数据类型
(3) stack space　　　　　　　　堆栈空间
(4) data structure　　　　　　　数据结构
(5) return address　　　　　　　返回地址

3. Identify the following to be True or False according to the text

T　F　T　F　T

4. Reading Comprehension

(1) c　either component values or component elements
(2) a　an atomic data type
(3) d　first-in/first-out
(4) b　first-in/last-out

2.2 System Software

1. Translate the following phrases into English

(1)文件存储空间　　　　　　file storage space
(2)操作系统　　　　　　　　operating system
(3)代码生成器　　　　　　　code generator
(4)词法分析器　　　　　　　lexical analyzer
(5)资源管理器　　　　　　　resource manager

2. Translate the following phrases into Chinese

(1) allocate resource　　　　分配资源
(2) single-task　　　　　　　单任务
(3) syntactic analyzer　　　　语法分析器
(4) code optimization　　　　代码优化
(5) semantic processor　　　语义处理器

3. Identify the following to be True or False according to the text

T　F　T　T　F

4. Reading Comprehension

(1) c　operating system
(2) c　job queue
(3) a　operating system
(4) d　grammar rules

2.3 C Language and C++ Language

1. Translate the following phrases into English

(1)高级语言　　　　　　　　high-level language
(2)源代码　　　　　　　　　source code
(3)库函数　　　　　　　　　library function
(4)面向过程的编程语言　　　procedure-oriented programming language
(5)数据隐藏　　　　　　　　data hiding

2. Translate the following phrases into Chinese

(1) structured programming language　　结构化编程语言
(2) machine language　　　　机器语言
(3) assembly language　　　　汇编语言
(4) object-oriented program　　面向对象程序
(5) functional detail　　　　　功能细节

3. Identify the following to be True or False according to the text

F　T　T　T　F

4. Reading Comprehension

(1)d　encapsulation, data hiding, inheritance, and polymorphism
(2)c　function
(3)a　1970s
(4)d　calling function

2.4　Java and Object-Oriented Programming

1. Translate the following phrases into English

(1)白皮书　　　　　　　　　　white paper
(2)存储器分配　　　　　　　　memory allocation
(3)虚函数　　　　　　　　　　virtual function
(4)传统的编译型语言　　　　　traditional compiled language
(5)本机代码　　　　　　　　　native code

2. Translate the following phrases into Chinese

(1)Web browser　　　　　　　　万维网浏览器
(2)portable application software　可移植应用软件
(3)source code　　　　　　　　源代码
(4)runtime exception handling　运行期异常处理
(5)Java virtual machine　　　　Java 虚拟机

3. Identify the following to be True or False according to the text

F　T　F　T　F

4. Reading Comprehension

(1)a　address
(2)a　properties
(3)b　software
(4)d　byte-code

2.5　Database Technologies

1. Translate the following phrases into English

(1)数据库管理系统　　　　　　database management system
(2)数据定义语言　　　　　　　data definition language
(3)数据字典　　　　　　　　　data dictionary
(4)物理细节　　　　　　　　　physical details

(5)完整性约束　　　　　　　　integrity constraint

2. Translate the following phrases into Chinese

(1)database system　　　　　　数据库系统
(2)database schema　　　　　　数据库模式
(3)storage manager　　　　　　存储管理器
(4)data warehouse　　　　　　　数据仓库
(5)decision-support　　　　　　决策支持

3. Identify the following to be True or False according to the text

F　　F　　T　　T　　T

4. Reading Comprehension

(1)d　　the user
(2)b　　file system
(3)a　　transaction
(4)c　　a database system

2.6　Software Engineering

1. Translate the following phrases into English

(1)内部数据结构　　　　　　　internal data structure
(2)仿真环境　　　　　　　　　simulated environment
(3)维护阶段　　　　　　　　　maintenance phase
(4)软件开发　　　　　　　　　software development
(5)集成测试　　　　　　　　　integration testing

2. Translate the following phrases into Chinese

(1)software engineering　　　　软件工程
(2)waterfall model　　　　　　瀑布模型
(3)software testing　　　　　　软件测试
(4)boundary condition　　　　　边界条件
(5)requirements specification　　需求说明

3. Identify the following to be True or False according to the text

T　　T　　F　　T　　F

4. Reading Comprehension

(1)c　　Integration testing
(2)a　　A function test
(3)d　　actual working environment
(4)b　　correcting problems

2.8.2 第 2 单元参考译文

2.1 数 据 结 构

2.1.1 课文 A

数据结构是一种数据类型，其值是由与某些结构有关的组成元素所构成的。它有一组在其值上的操作。此外，可能有一些操作是定义在其组成元素上的。由此我们可知：结构数据类型可以有定义在构成它的值之上的操作，也可以有定义在这些值的组成元素之上的操作。

数据类型

数据类型的本质是标识一组个体或目标所共有的特性，这些特性把该组个体作为可识别的种类。如果我们提供了一组可能的数据值及作用在这些数据值上的一组操作，那么这两者结合在一起就称为数据类型。

让我们看两种数据类型，我们称任何由原子值构成的数据类型为原子数据类型，通常我们倾向于把整数作为原子。那么，我们关心的仅仅是一个值所代表的单个量，而不是把整数看成一个在某些数字系统中的数字的集合。在许多程序设计语言和计算机体系结构中的整数类型是一个常用的原子数据类型。

人们可以将其值由某种结构相关的组成元素构成的数据类型称为结构化数据类型或数据结构。换句话说，这些数据类型的值是可分解的，因此我们必须知道它的内部结构。任何可分解的目标有两个必要的组成成分——必须具有组成元素和结构，即将这些元素相互关联或匹配的规则。

面向对象的数据结构

面向对象的程序设计是一种现代的软件开发方法，用这种方法设计的软件具有高的可靠性和灵活性。软件系统实现的复杂性包括信息表示的复杂性和对这些表示进行操作的算法复杂性。数据结构则是研究一些方法，这些方法用来表示对象、安全可靠的封装结构、研发采用这些表示的算法及测量因此而得到的系统的时空复杂度。面向对象的方法强调对象的作用及它们的属性和操作，这些构成了解决方案的核心。

在特定类中，从决定使用何种数据结构来表示对象的属性这一点来看，面向对象的方法中对抽象的强调，在软件开发过程中是非常重要的。抽象意味着隐藏不必要的细节。过程抽象或算法抽象是对算法隐藏细节的，允许算法在各个细节层次上可见或被描述。建立子程序是抽象的一个实例，子程序名描述了子过程的功能，子程序内部的代码表示了处理过程是如何完成的。

类似地，数据抽象隐藏了描述的细节。一个明显的例子是通过组合几种数据类型来构建新的数据类型，每种新类型描述了一些更复杂的对象类型的属性或组成。数据结构中面向对象的方法通过把对象类的表示整合将数据抽象和过程抽象组合在一起。

一旦选择了一个合适的抽象，就有一些选择来表示数据结构。在许多情况下，至少有一种静态表示和一种动态表示。在静态和动态表示中，典型的折中方法是介于针对存储空间的增加选择边界或非边界的表示以及和一些非边界表示有关联的时间需求之间。

在选择了抽象和表示后，就有各种不同的方法来封装数据结构。对封装的选择是另一种权衡，在如何使结构对用户有用和包怎样来操作用户的示例对象之间进行。封装对表示的完整性及与封装相关的时间、空间需求都有影响。一旦说明以后，一种或多种竞争的表示的方法将被执行，与解决的问题有关的结构、它的表示和封装将被评价。每种方法的时间和空间需求必须相对于系统需求和约束被衡量。

由于面向对象程序设计把对象当作数据结构，所以它不同于过程程序设计。结构化数据及其相关的操作可被封装在一个对象中，它可以被重复使用和易于升级、增加和替代。因此，它能直接降低维护费用和加快新系统的问世并提高其可扩充性。

2.1.2 课文 B

栈和队列

栈是一种数据类型，它的主要性质是由对其节点的插入与删除的管理规则来确定的，被删除或移去的节点只能是刚刚插入的，就是所谓具有后进先出（LIFO）行为或协议的结构。栈这种数据类型虽然简单，但并不影响其重要性，许多计算机系统的电路中都含有多个栈，并且含有操作硬件栈的机器指令。子程序的调用和返回顺序遵循栈的协议。算术表达式的计算通常是通过对栈的操作顺序来实现的。大多数袖珍计算器都是用栈模式来操作的，在学习计算机科学时，人们能看到许多栈的例子。

队列的例子在日常生活中经常出现且为我们所熟悉，在银行等待服务或在电影院门口等待买票的一队人、在交通灯前面等待通行的一长串汽车都是队列的例子。队列的主要特征是遵循先来先服务的原则。与栈最后插入的节点是最先删除或服务的不同，在队列中，最先插入的节点将最先被服务。这样的原则与社会生活中人们公平合理的想法是一致的。

队列的先进先出（FIFO）原则在计算机中有很多应用。例如，在多用户分时操作系统中，多个等待访问磁盘驱动器的输入/输出（I/O）请求就可以是一个队列。等待在计算机系统中运行的作业也同样形成一个队列，计算机将按照作业和 I/O 请求的先后次序进行服务，即先进先出。另外，还存在着一种重要的队列，这在日常生活中也是可以看到的，如在医院的急救室内。在危重病人多的情况下，医生必须首先抢救生命垂危的病人。

在计算机系统中，要求计算机系统服务的事件通常根据最重要的事件最先服务来处理，换句话说，是按服务优先级最高先进/先出队列（HPIFO）的原则，这种队列称为优先队列。优先队列并不按到达时间的先后决定服务的次序,而是按照优先级越高越优先服务的原则。

函数调用

当进行一个新函数调用时，所有局部于调用程序的变量都需要由系统存储起来，否则新函数将要重写调用程序的变量。而且调用程序的当前位置也必须保存，以便新函数知道它运行后返回何处。变量通常由编译器分配到机器寄存器，而且尤其是在涉及递归时，肯定会有冲突。

调用函数时，所有需要存储的重要信息如寄存器值和返回地址，都以抽象方式存于"一片纸"上，且放在一个堆的顶端。然后控制转向新函数，新函数可自由地用它的值替换寄存器的值。如果它再做其他函数调用，则可进行同样的步骤。当函数要返回时，先查看在堆顶的"纸片"，并恢复所有寄存器，然后进行返回跳转。

显然，所有这些工作都可用栈来完成，而且实际上在每种实现递归的程序语言中都是这样做的。保存的信息称为活动记录或栈框架。现实计算机中的栈常常由内存部分的高端向下延展，并且在很多系统中没有溢出检查。而且总是有可能，由于同时有太多运行的函数而使栈空间溢出，不用说，栈空间溢出是个致命的错误。

在对栈溢出不做检查的语言和系统中，程序可能没有合适的解释就崩溃了。在这些系统中，当栈太大时，常会发生奇怪事情，因为你的栈可以延伸进你的程序部分。它可能是主程序，也可能是部分数据，尤其是当有一个大型数组时。若栈延伸进你的程序，则程序可能会错误百出，并会产生一些毫无意义的指令，并且一经执行此指令，就会崩溃。如果栈延伸至数据，可能发生的情况是，当你向数据中写入某内容时，它将会破坏栈的信息——或许是返回地址——且程序将会试图返回到某个古怪地址并崩溃。

2.2 系 统 软 件

2.2.1 课文 A

操作系统是一种程序，它是用户与计算机硬件之间的接口。操作系统的目的是提供一种用户能执行程序的环境。然而，一般来说，操作系统没有一个完整恰当的定义。操作系统的存在是因为它们是解决建立可以使用的计算机系统问题的一种合理的方法。计算机系统的基本目标是执行用户程序和解决用户问题。计算机硬件是朝着这个目标而构建的。因为只有硬件的裸机不能使用，所以开发了应用程序。这些不同的程序要求某些共同的操作，例如控制 I/O 设备。这些控制和分配资源的共同功能合并到一个软件中，就形成了操作系统。

最初的操作系统是由计算机制造商在生产线上为他们的计算机产品开发的。当制造商推出另一种计算机或型号时，他们往往会开发出经过改进的且不同于原来的操作系统。这就意味着如果用户希望更换计算机，无论是更换为另一个制造商生产的计算机，还是更换为同一个制造商生产的另外一种型号的计算机时，都必须转换已有的程序使它们能在新的操作系统下运行。而今天，仅限于某个型号的操作系统已经被淘汰，取而代之的是能运行于指定制造商生产的任何型号的操作系统上。

研究操作系统有很多重要的理由，最突出的是：

◆ 用户为了完成自己的任务必须与操作系统打交道,因为操作系统是用户与计算机之间的主要接口；

◆ 大多数计算机安装的主要决策是选择操作系统及其选项；

◆ 操作系统中所发现的许多概念和技巧一般可以应用到其他领域中；

◆ 为了某些特殊应用，人们可能不得不设计自己的操作系统，或者对已有的系统进行改进。

一个操作系统与一个政府类似。计算机系统的基本资源由硬件、软件与数据提供。操作系统为计算机系统的操作提供正确使用这些资源的方法。像一个政府一样，操作系统只不过是提供一个环境，在该环境中其他程序能发挥作用。

我们可以把操作系统看成一个资源分配器。计算机系统有很多的资源用来解决一个问题：CPU 的时间、存储空间、文件存储空间、输入/输出设备等。操作系统作为资源的管理者，为满足特定程序和用户的任务需要而分配资源。因为有许多资源请求方面可能存在冲突，操作系统必须决定给哪些请求分配资源以便计算机系统能合理而有效地运行。

操作系统有单任务和多任务之分。早期的许多单任务操作系统在同一时间只能运行一个进程。例如，当计算机打印文件时，它就不能开始运行另一个进程，或者不能响应新的命令，直到打印完成为止。

所有现代操作系统都是多任务的，同时能运行几个进程。因为大部分计算机中仅有一个 CPU，所以多任务操作系统让人产生 CPU 能同时运行几个进程的错觉。产生这种错觉的最常用机制是时间片多任务处理，它以每个进程各自运行固定的一段时间的方式来实现。如果一个进程在分派的时间内没有完成，它就被挂起，另一个进程接着运行。这种进程交换称为任务切换。操作系统实行"簿记"法以保存被挂起的进程状态。它同样有一种机制，称为调度程序，由它决定下一时刻将运行哪个进程。为了把感觉到的延迟减到最小，调度程序运行短进程非常迅速。由于用户的时间感觉比计算机的处理速度要慢得多，所以几个进程看起来是同时执行的。

操作系统的一个非常重要的职责是调度计算机系统将要处理的作业。这是作业管理功能的主要任务之一。操作系统建立程序处理的顺序，并定义了具体作业执行的次序。术语"作业队列"常用于描述等待执行的作业序列，操作系统在排列作业队列时将权衡各方面因素，包括当前正在处理哪些作业，正在使用哪些系统资源，需要哪些资源来处理后面的程序，与其他任务相比该作业的优先级及系统应响应的一些特殊处理要求等。操作软件应能评估这些因素并控制作业处理的顺序。

2.2.2 课文 B

编译程序指将用高级语言书写的文本转换成等价的用低级汇编语言或机器语言书写的文本。我们将大多数现存的编译程序分解成七个公用部分，观察它们之间的关系。这七个部分的内容如下：

- ◆ 扫描程序
- ◆ 语法分析程序
- ◆ 中间代码生成程序
- ◆ 语义处理程序
- ◆ 优化程序
- ◆ 代码生成程序
- ◆ 表格

不是所有的编译程序都具有这七个由彼此独立的实体构成的模块，例如，扫描程序可以合并到语法分析程序中去，或者语义处理程序可以合并到语法分析程序中去，等等。但是，从概念上看，把任一编译程序分解成这七个部分是较容易的。

扫描程序在有些书中称为词法分析程序，当一个程序首次输入编译程序中时，它只是一个长符号流，也许分析成与输入介质有关的行或记录。扫描程序是把源程序的这种外部形式转换成更适用于编译程序其他部分操作的内部形式。

扫描程序的作用如下：
- 指出程序中的基本词法单元，它们被称为标记；
- 删除无关的空格、回车符及其他与输入介质有关的符号；
- 删除注释行；
- 报告扫描器发现的错误。

典型的扫描程序对原始形式的文本进行一次扫描，完成四个任务，然后采用内部格式，以一次一个标记或以一个大的标记文件形式输出给语法分析程序。扫描程序通常是逐个字符地审查文本。

每种程序设计语言都有自己的语法规则集，用于说明本语言编制程序的正确形式。分析器或语法分析程序接收扫描器的输出，即标记和证实所编译的源程序满足本语言的语法规则。由语法分析器输出的语法树被转换成用中间代码所写的某种形式的"程序"，这种形式与源文本相比更接近于汇编语言，它比汇编（机器）代码更便于被编译程序以后处理。

代码生成器直接把中间代码转换成目标语言。然而，通常在语义处理器和代码生成器之间插入一个成分，即优化程序（代码优化）。如果中间代码直接由代码生成器转换成汇编代码，则产生的目标代码在执行时间和存储空间方面的效率往往不高。如果一个编译器要产生高效的代码，正如大多数商业编译器那样，则它必须包含一个特定设计的模块以综合改进代码的时间和空间特性。优化程序将原代码改为更有效的形式。

我们把优化技术分成两类：一类是在源程序（它的内部形式）上完成的优化，因此它们是不依赖于目标语言的；另一类是在目标程序级上完成的优化。

代码生成器将优化器产生的中间代码转换为汇编或机器语言代码（目标程序）；显然，代码生成是高度依赖于机器的。因此，当目标机改变时，代码生成器也必须随之改变。另外，其他部分对产生代码的机器的改变是不敏感的。然而，通常没有这样的编译器模块，可以在目标机改变时，一点也不改变。

2.3 C 语言与 C++语言

2.3.1 课文 A

C 是一种通用的结构化编程语言，它的指令是由一些类似代数表达式的项加上一些英文关键字如 if、else、for、do 和 while 而组成的。从这方面讲，C 语言类似于其他高级结构式编程语言，但 C 语言还另有一些特点。这些特点使它能在较低层次上应用，因而能弥补机器语言和惯用的高级语言之间的差距。这种灵活性使 C 语言可以用于系统编程（例如编写操作系统），也可以用于应用编程（例如，编写解数学方程式的复杂系统的程序或者编写给顾客开账单的程序）。

C 语言的研制始于 20 世纪 70 年代初期。C 语言也许最适宜称为"中级语言"。像真正的高级语言一样，一个 C 语句与编译后的机器语言指令的关系是一对多的关系。因此，像 C 语言这样的语言编程手段远远超过低级的汇编语言。然而，与大多数高级语言相比，C 语言有一个很小的结构集。另外，与大多数高级语言不同，C 语言可使操作者很容易地做由汇编语言执行的工作（如位与指针的操作）。

结构化语言

尽管块结构语言这个术语从学术上考虑并不能严格地应用于 C 语言，但 C 语言是那个语言族的非正式成员。块结构语言的显著特征是代码和数据的区域化，这意味着语言能够把完成某个具体任务所必需的所有信息及指令和程序的其余部分分割开并隐藏起来。区域化一般是利用含有局部或临时变量的子程序获得的。用这种方法，用户可以编写各种子程序，在这些子程序内部发生的事件对程序的其他部分不会产生副作用。过度使用整个程序都理解的全局变量会在程序中引起错误及意想不到的副作用。在 C 语言中，所有的子程序都是独立的函数。

函数是 C 语言的组成模块，程序的所有活动都发生在函数中，它们允许用户分别定义和编写程序中的各个特定的任务。使用局部变量的函数调试过后，用户就可以依赖它在各种情况下良好地工作，不会对程序的其他部分产生副作用。在某个函数中定义的全部变量只有那个函数才认识。

C 程序由一系列函数组成，程序的执行必须以称为 main() 的函数开始，包含在程序中的其他函数由程序员命名。当函数名在程序中以一个语句的形式出现时，程序的执行转到那个函数，当被调用函数执行完时，将把一个结果返回给调用函数。如果一个函数调用另一个函数，那么第二个函数则是嵌套在第一个里的。在许多情况下，当一个函数被调用时，有信息传送给它，这个信息包含在函数名后一对括号内。被调用函数也将返回一个值给调用函数。

特点

C 语言的特点是它能写出很简明的源程序，其部分原因在于该语言包括大量的运算符。它的指令集相对较小，不过实际的实现包括大量的库函数，这些库函数增强了基本指令。此外，该语言鼓励用户编写他们自己的附加库函数。因此，用户可以方便地扩展 C 语言的属性和能力。

C 语言的编译程序普遍适用于各种容量的计算机，并且 C 语言的解释程序正变得越来越普通。编译程序是紧凑的，它们生成的目标程序较之由其他高级语言编译所得的程序要短小且高效。尽管在开发新程序时，使用解释程序较容易，但效率低。许多程序员都从解释程序开始，然后一旦程序调式完毕（即一旦编程错误全部被排除后）便转用编译程序。

C 语言的另一个重要特点是它的程序具有高度的可移植性，与其他高级语言相比，更是如此。其原因是 C 语言把大多数与计算机有关的特性都归进了它的库函数。因此，每个版本的 C 语言都伴有它自己的库函数集，这些库函数集是按主机的特点而编写的。这些库函数是相对标准化的。一般而言，不同版本的 C 语言访问库函数的方法都一样。因此，对大多数 C 程序可不做任何修改或做很少改动就能在许多不同的计算机上被处理。

2.3.2 课文 B

C++ 是一种通用的计算机编程语言，具有高级和低级处理的能力。它是静态输入、自由格式、多范例、编译程序支持的过程编程、数据抽象、面向对象编程及通用编程的语言。C++ 是 C 的加强版。C++ 包括 C 的一部分并且增加了对面向对象编程的支持。C 是一种面向过程的编程语言。C++具有更多的改进和特性。C 和 C++的基本语法和语义是相同的。如果你对 C 较熟悉，那么你可以很快地学会 C++编程。C++拥有在 C 中定义的类似的数据类型、操作符及其他工具，能直接适用于计算机体系结构。

C++完全支持面向对象编程，它包括以下四个面向对象的开发工具：封装，数据隐藏，继承和多态。尽管 C++是 C 的超集，而且任何合法的 C 程序也都是合法的 C++程序，但从 C 跳到 C++的意义是非常重大的。由于 C 程序员都能轻松地进入 C++世界，因此，多年来 C++从其与 C 的关系中获益匪浅。然而，如果想真正精通 C++，你会发现你不得不放弃很多已有的观念，并学习一种新的分析和解决编程问题的方法。

面向对象的程序设计是一种程序设计技术，使你能把一些概念看成各种各样的对象。通过使用对象，你能表示要被执行的任务、它们之间的相互作用和必须观察的某些给定的条件。一种数据结构经常形成某个对象的基础；因此，在 C 或 C++中，结构类型能形成某种基本对象。与对象的通信能通过使用消息来完成。消息的使用类似于在面向过程的程序中对函数的调用。当某对象收到一个消息时，包含在该对象内的一些方法做出响应。方法类似于面向过程程序设计的函数。然而，方法是对象的一部分。

C++通过创建称为类的用户定义类型而支持封装和数据隐藏的属性。类一经创建，就可作为一个完全封装的实体——一个完整的单元。类的实际内部工作都应该被隐藏。一个定义好类的用户只需知道怎样使用它就可以了，而不必知道类是如何工作的。

C++类是 C 结构的扩展。由于结构与类的唯一区别在于结构成员的默认访问权限是公共的，而类成员的默认访问权限是私有的，因此你可以使用关键字类或结构来定义相同的类。C++类是对 C 和 C++结构类型的扩充，并且形成了面向对象程序设计所需要的抽象数据类型。类能包含紧密相关的一些项，它们共享一些属性。更正式地说，对象只不过是类的实例。最终，应该出现包含很多对象类型的类库，你能使用这些对象类型的实例去组织程序代码。

典型地，一个对象的描述是一个 C++类的一部分，并且包括该对象内部结构的描述、该对象如何与其他对象相关，以及把该对象的功能细节和该类的外部相隔离的某种形式的保护。C++类结构实现了所有这些功能。

在一个 C++类中，你使用私有的、公共的或受保护的描述符来控制对象的功能细节。在面向对象的程序设计中，公共部分一般用于接口信息（方法），使得该类可在各应用中重用。如果数据或方法被包含在公共部分，则它们在该类外部也可用。类的私有部分把数据或方法的可用性局限于该类本身。包含数据或方法的受保护部分被局限于该类和任何派生子类。

2.4　Java 与面向对象程序设计

2.4.1　课文 A

Java 已经变得极为流行。Java 的快速崛起和广为接受可归因于其设计与编辑特点，特别是它承诺：程序一次编写，可在任何地方运行。正如 Sun 公司的 Java 语言白皮书所说，Java 是一种简单的、面向对象的、分布式的、解释型的、健壮的、安全的、体系结构中立的、可移植的、高性能的、多线程的和动态的语言。简言之，Java 环境可用来开发能在任何计算平台上运行的应用软件。它实际上是一种非常基本且结构紧凑的技术，而它对万维网及商业的整个影响已可同电子表格对 PC 的影响相比拟。

Java 比 C++这种流行的面向对象程序设计语言要容易些。在 Java 之前，C++曾是占支配地位的软件开发语言。Java 部分仿照了 C++，但大大简化和改进了。例如，指针和多重继承常常使编程变得复杂。Java 用一种称为接口的简单语言结构取代了 C++中的多重继承，并取消了指针。

Java 采用自动的存储分配和垃圾收集，而 C++要求程序员分配存储器和收集垃圾。此外，就这样一种功能强大的语言而言，语言结构的数量算小的。简洁的语法使 Java 程序易于编写和读取。

- **Java 是一种面向对象语言**

尽管不少面向对象语言在开始时完全是过程语言，但 Java 从一开始就是要面向对象的。面向对象编程（OOP）是一种流行的编程方法，它正在取代传统的过程编程技术。

使用过程编程语言开发的软件系统基于过程范式。面向对象编程使用对象来对真实世界进行建模。世界上的任何东西都可作为一个对象来建模。一个圆是一个对象，一个人是一个对象，一个 Windows 图标也是一个对象，甚至一笔贷款也可视为一个对象。Java 程序是面向对象的，因为用 Java 进行的编程所围绕的是创建对象、操纵对象及使对象协作。

软件开发的一个中心问题是如何重复使用代码。面向对象编程凭借封装、继承和多态性提供了很大的灵活性、模块性、明确性和可重复性。多年来，面向对象技术被认为是精英技术，要求在培训和基础设施方面进行可观的投入。Java 帮助面向对象技术进入了计算领域的主流，其简洁明了的语法使程序易于编写和读取。就设计和开发应用程序而言，Java 程序具有相当的表现力。

- **Java 是一种分布式语言**

分布式计算涉及数台计算机在网络上协作。Java 旨在使分布式计算变得容易。由于联网能力一开始即被结合进 Java，所以所有编写网络程序宛如向一个文件发送或从一个文件接收数据。如果你下载一个 Java 小程序（一种特殊的程序）并在你的计算机上运行，它不会破坏你的系统，因为 Java 实施几种安全机制来保护你的系统免于杂乱程序造成的危害。

- **Java 是一种解释型语言**

运行 Java 程序需要解释器。程序被编译成称为字节码的 Java 虚拟机代码。字节码独立于机器，可在有 Java 解释器的任何机器上运行，而 Java 解释器是 Java 虚拟机的组成部分。

大多数编译器，包括 C++编译器，将高级语言程序翻译成机器代码。这种代码只能在本机上运行。如果在其他机器上运行程序，则得在本机上重新编译程序。例如，如果在 Windows 中编译 C++程序，由编译器生成的可执行代码只能在 Windows 平台上运行。就 Java 而言，你只需编译源代码一次，而由 Java 编译器生成的字节码可在任何有 Java 解释器的平台上运行。Java 解释器将字节码翻译成目标机的机器语言。

- **Java 是一种健壮的语言**

健壮意指可靠。没有哪种程序设计语言能够确保安全可靠。Java 非常重视对可能存在的错误进行早期检查，Java 编译器可检测出许多在其他语言中执行时才会首次暴露出来的问题。Java 取消了在其他语言中发现的某些类型的易出错编程结构。例如，它不支持指针，从而消除了覆写内存和破坏数据的可能性。

Java 拥有运行期异常处理的功能，用于为健壮性提供编程支持。Java 迫使程序员编写用于处理异常的代码。Java 可捕捉异常情况并对其做出反应，以便在发生运行期错误时，程序能够继续正常运行，并从容终止。

- Java 是一种体系结构中立的语言

Java 是一种解释型语言。这个特征使 Java 能做到体系结构中立，或者换个说法，独立于平台。凭借 Java 虚拟机，你可以编写能在任何平台上运行的程序。

Java 最初的成功源自它的万维网编程能力。你可以从一个万维网浏览器运行 Java 小程序，但 Java 不只是用于编写万维网小程序的。你还可以使用 Java 解释器从操作系统直接运行独立的 Java 应用程序。今天，软件供应商通常会将同一种产品开发成多个版本，以便在不同的平台上运行。开发者使用 Java，则只需编写一个版本即可，而该版本可在每个平台上运行。

由于 Java 是体系结构中立的，所以 Java 程序是可移植的。它们可在任何平台上运行而不需重新编译。

2.4.2 课文 B
面向对象程序设计（OOP）

面向对象程序设计的基本原则是：一个计算机程序是由一系列独立单元的集合组成的，或者说是由那些像子程序一样起作用的对象组成的。为了使整个计算程序可以运行，每个对象都要有能力接收信息、处理数据及发送信息给其他的对象。简而言之，对象之间通过自有的功能（或方法）及数据可以相互作用。

通过这种方式，消息可被正确地用一块或更多量的代码以一种无缝的方式进行处理。面向对象程序设计主张的是：它能够在简单的、按部就班的程序设计方式（即计算机领域内所谓的命令式程序设计或结构化程序设计）中增加更多的灵活性。

OOP 的提出者同时也指出，对那些初次学习计算机程序设计的人来说，OOP 比以往的方式方法更加直观和简单易学。在 OOP 方法里，对象通常是单一的、独立封装的、易于辨认的。这种模块性质决定了程序块与真正需要解决的问题之间相协调，进而也与真实的世界相适应。面向对象程序设计通常从问题产生的条件陈述开始，然后通过嵌入对象或名词变量、动词方法和形容词属性，就形成了一个良好的建立在程序模式上的解决问题的框架结构。这时就要求你能学习使用面向对象编程语言来设计程序。

面向对象程序设计中的继承使得一个类能继承某对象类的一些性质。父类用作派生类的模式，且能以几种方式被改变。如果某个对象从多个父类继承其属性，便称为多继承。继承是一个重要概念，因为它使得无须对代码做大的改变就能重用类定义。继承鼓励重用代码，因为子类是对父类的扩充。

与类层次结构相关的另一个重要的面向对象概念是，公共消息能被发送到各个父类对象和所有派生子类对象。按正式的术语，这称为多态性。

多态性使每个子类对象能以一种对其定义来说适当的方式对消息格式做出响应。试设想收集数据的一个类层次结构。父类可能负责收集某个个体的姓名、社会安全号、职业和雇用年数，那么你能使用子类来决定根据职业将添加什么附加信息。一种情况是一个管理职位会包括年薪，而另一种情况是销售员职位会包括小时工资和回扣信息。因此，父类收集一切子类公共的通用信息，而子类收集与特定工作描述相关的附加信息。多态性使得公共的数据收集消息能被发送到每个类。父类和子类两者都对该消息以恰当的方式做出响应。

多态性赋予对象这种能力，当对象的精确类型还未知时响应来自例行程序的消息。在C++中这能力是迟绑定的结果。使用迟绑定，地址在运行时刻动态地确定，而不是如同传统的编译型语言在编译时刻静态地确定。该静态的方法往往称为早绑定。函数名被替换为存储地址。你使用虚函数来完成迟绑定。一个父类，当随后的各派生类通过重定义一个函数的实现而重载该函数时，便在其中定义了虚函数。当你使用虚函数时，消息不是直接传给对象，而是作为指向对象的指针传送。

虚函数利用了地址信息表，该表在运行时刻通过使用构造符初始化。每当创建它的类的一个对象时调用一个构造符。构造符的工作是把虚函数与地址信息表链接，在编译进行期间，虚函数的地址是未知的；相反，给出的是地址表中将包含该函数地址的位置。

2.5 数据库技术

2.5.1 课文 A

数据库管理系统（DBMS）由一个互相关联的数据的集合和一组用以访问这些数据的程序组成，这个数据集合通常称为数据库，其中包含了关于某个企业的信息。DBMS 的主要目标是要提供一个可以方便地、有效地检索和存储数据库信息的环境。

设计数据库系统的目的是为了管理大量信息。对数据的管理既涉及信息存储结构的定义，又涉及信息操作机制的提供。另外，数据库系统还必须保证所存储信息的安全性，即使在系统崩溃或有人企图越权访问时也应保证信息的安全性。如果数据被多用户共享，那么系统还必须设法避免可能产生的异常结果。对大多数组织而言，信息都非常重要，这决定了数据库的价值，并使得大量的用于有效数据管理的概念和技术得到发展。

数据库系统所使用的存储结构和访问方式通过一系列特殊的 DDL 语句来定义，这种特殊的 DDL 语句称为数据存储定义语言。这些语句的编译结果是一系列用来描述数据库模式实现细节的指令，这些实现细节对用户来说通常是不可见的。数据库模式也是通过 DDL 说明的。DDL 语句的编译结果产生存储在一个特殊文件中的一系列表，称为数据字典或数据目录。数据字典是一个包含元数据的文件，元数据是关于数据的数据。在数据库系统中，在读取和修改实际数据前总要先查询该文件。

事务管理

事务是数据库应用中完成单一逻辑功能的操作集合。每个事务是一个既具原子性又具一致性的单元。因此，我们要求事务不违反任何的数据库一致性约束，也就是说，如果事务启动时数据库是一致的，那么当这个事务成功完成时数据库也应该是一致的。但是，在事务执行过程中，必要时允许暂时的不一致。这种暂时的不一致尽管是必需的，但在故障发生时，很可能导致问题的产生。

存储管理

数据库常常需要大量存储空间。公司数据库的大小是用 gigabyte 来计算的,最大的甚至需要用 terabyte 来计算。1gigabyte 等于 1000megabyte(10^6 字节),1terabyte 等于 100 万 megabyte(10^{12} 字节)。由于计算机主存不可能存储这么多信息,因而信息被存储在磁盘上,需要时信息在主存和磁盘间移动。由于同中央处理器的速度相比,数据出入磁盘的速度很慢,数据库系统对数据的组织必须满足使磁盘和主存间数据移动的需求最小化的要求。

数据库系统的目标是简化和辅助数据访问。高层视图有助于实现这样的目标。系统用户可以不受系统实现的物理细节所带来的不必要的负担所累。但是,决定用户对数据库系统是否满意的一个主要因素是系统性能。如果一个请求的响应速度太慢,系统的价值就会下降。系统性能取决于用来表示数据库中数据结构的高效性,以及系统对这样的数据结构进行操作的高效性。正如计算机系统中其他方面的情况一样,不仅要在时间与空间两者间进行权衡,还要在不同操作的效率间进行权衡。

存储管理器是在数据库中存储的低层数据与应用程序及向系统提交的查询之间提供接口的程序模块。存储管理器应负责与文件管理器的交互。原始数据通过文件系统存储在磁盘上,文件系统通常由传统的操作系统提供。存储管理器将不同的 DML 语句翻译成低层文件系统命令,因此,存储管理器负责数据库中数据的存储、检索和更新。

2.5.2　课文 B

数据仓库是从多数据源收集来的信息的仓储(或转储)库,它在一个地点以统一的模式存储数据。数据一旦收集进来,便被长期保存,以支持对历史数据的访问。因此,数据仓库提供给用户一个统一的数据接口,使得决策支持查询的书写更为容易。另外,通过从数据仓库访问用于决策支持的信息,决策制定者保证了联机事务处理系统不会受决策支持工作负载的影响。

图 2-1 给出了一个典型的数据仓库结构,并且表示出了数据的收集、数据的存储及查询和数据分析支持。构造数据仓库面临的问题如下:

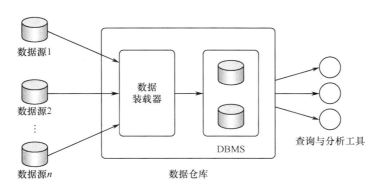

图 2-1　数据仓库结构

- 何时及如何收集数据。在数据收集的源驱动体系结构中,数据源要么连续地在事务处理发生时传送新信息;要么阶段性地,例如每天晚上传送新信息。在目标驱动体系结构中，数据仓库阶段性地向源发送对新数据的请求。

除非对源的更新通过两阶段提交在数据仓库中做了复制，否则数据仓库不可能总是与源同步。两阶段提交通常因开销太大而不被采用，所以数据仓库常会保留稍微有点过时的数据。但这对于决策支持系统来说通常不是问题。

- 采用什么模式。各自独立构造的数据源可能具有不同的模式。事实上，它们甚至可能使用不同的数据模型。数据仓库的部分任务就是做模式集成，并且在存储数据前将数据按集成的模式转化。因此，存储在数据仓库中的数据不仅仅是源中数据的拷贝，同时它们也可被认为是源中数据所存储的视图（或实体化的视图）。
- 如何传播更新。数据源中关系的更新必须被传至数据仓库。如果数据仓库中的关系与数据源中的一模一样，传播过程则是直接的。
- 汇总什么数据。由事务处理系统产生的原始数据可能太大以致不能联机存储。但是，我们可以通过维护对关系做聚集而得到的汇总数据回答很多查询，而不必维护整个关系。例如，我们不是存储每件服装的销售数据，而是按类存储服装的销售总额。

2.6 软件工程

2.6.1 课文 A

软件工程的早期方法坚持要严格地遵守分析、设计、实现及测试的顺序。因此，软件工程师坚持应在设计之前进行完整的系统分析，同样，设计应该在实现之前完成。这就形成了一个现在称为瀑布模型的开发过程，这是对开发过程只允许以一个方向进行的事实的模拟。

第一阶段是需求定义阶段，是指系统期望的需求分析，即描述功能特征和操作细节的阶段。如果系统是一个在竞争的市场上销售的通用产品，那么这个分析将会包括一个广泛的调查来发现潜在客户的需求。但是，如果系统是为特殊用户设计的，那么这个过程就是一个更专业的调查。当潜在用户的要求被确定之后，要将这些要求汇编成新系统必须满足的需求。这些需求是从应用的角度来表述的，而不是用技术术语来表达的。

一种需求可能是对数据的存取必须限制于有权限的人员。另一种需求可能是当最后一个工作日结束时，数据必须反映目前的清单状态，或者可能是在计算机屏幕上的数据组织必须按照目前使用的纸质形式的格式来显示。系统的需求被确定以后，它们就转化为更多的系统技术说明书。在需求阶段，一个错误的功能说明会导致不满足需求的设计和实现。如果没有检查就让错误发展下去，那么到了测试阶段就会花大量的财力去修正这个错误（包括重新设计和重新实现）。

第二阶段是设计阶段。设计关注这个系统应该如何实现目标。正是通过设计建立了软件系统的结构。这个阶段的输入是一份（经过调试和确认过的）需求文档，输出是以某种适当形式表示出的设计，如"伪代码"。确认设计阶段的正确性是非常重要的。需求文档中的每个需求都必须有相应的设计片段与之相符合。正规的验证虽然可以达到一定的程度，但却是极其困难的。更多的是整个的设计团队、管理者甚至是客户的非正式的校阅。

大型软件系统最好的结构是模块化系统，这是一条被充分证实的原则。确实，正是借助模块化的分解方法，大型系统的实现才成为可能。没有这样的分解，在大型系统实现过程中所需要的技术细节可能会超过一个人的理解能力。然而，有了这种模块化设计，仅仅需要熟悉与在考虑中的模块相关的细节。同样，模块化设计对未来的维护是有益的，因为它允许对基本的模块进行修改。

第三阶段是软件实现阶段，它是对第二阶段设计开发的实际编码阶段。设计必须被翻译成机器可识别的形式。编码阶段的任务就是做这样的工作。如果设计做得详细，编码就能由机器完成。这个阶段具有很大的诱惑力，很多鲁莽的程序员没有经过前两个阶段的充足准备就跳到了软件实现阶段。结果是，对需求关系没有完全弄清楚，设计也有缺陷。软件实现进行得很盲目，结果是出现越来越多的问题。

第四阶段是软件测试阶段。测试与实现紧密联系，因为对系统中的每个模块都要在实现的过程中进行正常的测试。在一个大系统的开发中，测试包括若干个阶段。首先，每个程序模块作为一个单独的程序进行测试，通常与系统中的其他程序分开。这种测试被定义为模块测试或单元测试。在任何可能的时候，单元测试是在控制环境下进行的，这是为了让测试小组能提供被测试模块先前确定的数据并观察产生的输出数据。另外，测试小组要检查内部数据结构、逻辑，以及输入和输出数据的边界条件。

在对模块集进行单元测试后，下一步是确保各模块间接口的定义和处理适当。综合测试是确定系统里的部件像程序设计和系统设计说明书中所描述的那样一起工作的过程。一旦我们确定信息按照设计要求在模块间传递，我们测试系统来确保它具有预期的功能。功能测试通过对系统进行评估来测定需求说明书中描述的功能能否在整个系统中都实现。

2.6.2 课文 B

无论我们写的程序有多好，我们显然都应该检查各种各样可能发生的错误来保证模块编码正确。许多程序员将测试看成他们所编程序能正确执行的证明。然而，证实正确性的观点实际上与测试的真正含义恰恰相反。我们测试程序是为了证明错误的存在。因为我们的目的是发现错误，只有当错误被发现了，我们才认为测试成功。一旦错误被发现，调试或纠错是确定引起错误的原因的过程，并且为了让错误不再存在要对系统做出修改。

任何有经验的程序员都会在心里测试生成的每行语句并于任何正式测试阶段之前在心里已经模拟了任何模块的执行过程。一个"成功"的测试只意味着在特定的测试环境下没有发现错误，但并不能说明在其他环境下也没有错误。理论上，测试表明程序正确的唯一方法是尝试所有可能的情况（称为穷举测试），这种情况对于最简单的程序在技术上都是不可能实现的。例如，假设我们编写了一个程序来计算一次考试的平均成绩。穷举测试要求测试所有可能的分数和班级人数的组合；完成这项测试可能需要很多年。

回忆一下，对需求下过的两种定义：一种是客户的术语，另一种是一系列软件和硬件的需求。功能测试是将建立的系统同软硬件需求中描述的功能进行比较。此外，执行测试是将系统与软硬件需求中的其他内容进行比较。如果测试是在客户的真实工作环境中执行的，一个成功的测试则会产生一个合法的系统。但是，如果测试是在一个模拟环境中执行的，则产生的系统只是验证系统。

当执行测试完成后，作为开发人员的我们要确保系统功能与我们对系统描述的理解相一致。下一步就是与客户商议确保系统的工作情况与客户的期望相一致。我们同客户一起执行验收测试，再次检查系统是否满足客户的需求。当验收测试完成后，已认可的系统被安装在它将被应用的环境中，然后执行最后的安装测试确保系统仍具有它应有的功能。

系统交付后，维护就开始了。维护工作包括对软件产品的改进，使产品适应新环境及修正问题。软件产品的改进包括提供新功能、用户显示及交互方式的改善，外部文件及内部文档的改善或者提高系统执行特性。使软件适应新环境应包括将软件应用到不同的机型上。例如，对软件进行修改使其适应一种新的电信协议或附加盘驱。问题修正包括对软件的修改及再次确认以改正错误。对一些错误需要及时注意，对有些错误可以根据时间表定期对其进行修改，对其余的只需知道而不必修改。

Unit 3   Multimedia and Its Applications

3.1   Computer Graphics

3.1.1   Text A

Computer graphics is a wonderful invention in the field of computers. It is used in diverse areas such as displaying the results of engineering and scientific computations and visualization, producing television commercials and feature films, simulation and analysis of real world problems, computer aided design, graphical user interfaces the communication bandwidth between humans and machines, etc.[1] The art of creating pictures with a computer has got numerous applications, that it is of great importance to explore the intrinsic of the world of computer graphics.

Graphics Primitives

A primitive is a graphics object that is essential for the creation or construction of complex images. Fortunately, graphics is constructed from three basic elements, as opposed to the great variety of graphics applications. The most basic of these elemental structures is the pixel, short for picture element.

A pixel is a point of light. It is just one tiny dot on the raster displays. Though it has no structure, it is definitely a building block and hence it can be considered as the graphics primitive. The resolution of CRT is related to the dot size, the diameter of a single dot. A resolution of 100 dots /inch implies a dot size of 0.01 inch. However, in reality, pixels are more elliptic than circle. The shape of a pixel purely depends upon the characteristics of the visual display unit. The ratio of the distance between the centers of two adjacent horizontal pixels to that of the vertical ones is called the pixel ratio.

Lines, especially straight lines, constitute an important building block of computer images. For example, line is the basic building block of line graphs, bar and pie charts, two and three-dimensional graphs of mathematical functions, engineering drawings and architectural plans. In computer graphics, straight line is so basic in creating images that we call it a graphics primitive. Straight lines can be developed in two different ways. A structural method determines which pixels should be set before drawing the line, a conditional method tests certain conditions to find which pixel should be set next.[2]

A polygon, even though generally constructed from straight lines, is an important graphics primitive. [3] So often we want to handle polygon as a single entity, as images of objects from the real world consist in large part of polygons. A polygon is a closed area of image bounded by straight or curved lines and filled with one solid color. Since images are two dimensional, a polygon is a plane figure.

Implementing a polygon as a graphics primitive is natural and helpful. We can define polygon as an image which consists of a finite ordered set of straight boundaries called edges. Alternately, the polygon can be defined by an ordered sequence of vertices, the corners of the polygon. The edges of the polygon are then obtained by traversing the vertices in the given order. The edge list is sufficient for wire frame drawings. Two consecutive vertices define one edge. We close the polygon by connecting the vertex to the first.

Output Primitives

Typically, graphics programming packages provide functions to describe a scene in terms of these basic geometric structures, referred to as output primitives, and to group sets of output primitives into more complex structures. Each output primitive is specified with input coordinate data and other information about the way that the object is to be displayed. Points and straight line segments are the simplest geometric components of pictures. Additional output primitives that can be used to construct a picture include circles and other conic sections, quadric surfaces, spline curves and surfaces, polygon color areas, and character strings.

Point plotting is accomplished by converting a single coordinate position furnished by an application program into appropriate operations for the output device in use. With a CRT monitor, for example, the electron beam is turned on to illuminate the screen phosphor at the selected location. How the electron beam is positioned depends on the display technology.

Line drawing is accomplished by calculating intermediate positions along the line path between two specified endpoint positions. An output device is then directed to fill in these positions between the endpoints. For analog devices, such as a vector pen plotter or a random-scan display, a straight line can be drawn smoothly from one endpoint to the other. Linearly varying horizontal and vertical deflection voltages are generated that are proportional to the required changes in the x and y directions to produce the smooth line.

Digital devices display a straight line segment by plotting discrete points between the two endpoints. Discrete coordinate positions along the line path are calculated from the equation of the line. For a raster video display, the line color (intensity) is then loaded into the frame buffer at the corresponding pixel coordinates. Reading from the frame buffer, the video controller then plots the screen pixels. Screen locations are referenced with integer values. This rounding of coordinate values to integers causes lines to be displayed with a stair step appearance (the zigzag). The characteristic stair step shape of raster lines is particularly noticeable on systems with low resolution, and we can improve their appearance somewhat by displaying them on high-resolution systems.[4] More effective techniques for smoothing raster lines are based on adjusting pixel intensities along the line paths.

Key Words

adjust	调整，校正
alternately	轮流地，交替地
appearance	外观，出场
appropriate	适当的，恰当的
architectural	建筑上的，建筑学的
conic	圆锥曲面
consecutive	连续的，依次相续的
corner	角落，角，拐角
curve	弧线，曲线，弯曲
deflection	偏斜，偏角，偏转
dimensional	维的，尺寸的
discrete	分离的，不相关的，非连续
diverse	不同的，多种多样的，变化多的
elliptic	椭圆形的
endpoint	终点
feature	特征，特点
fortunately	幸运地
intensity	强烈，强度，力度
intrinsic	本质的，真正的
numerous	很多的，多数的
phosphor	荧光
pixel	像素
polygon	多边形
proportional	比例的，成比例的，相称的
purely	完全地，十足地
quadric	二次曲面
raster	光栅
ratio	比率，系数，比值
scene	场景，布景
smoothly	平滑地，流畅地
solid	可靠的，实心的，固体的
spline	齿条，花键，方栓
straight	直，整齐
vertex（复数 vertices）	顶点，最高点
visualization	可见性，形象化
zigzag	锯齿

Notes

[1] It is used in diverse areas such as displaying the results of engineering and scientific computations and visualization, producing television commercials and feature films, simulation and analysis of real world problems, computer aided design, graphical user interfaces the communication bandwidth between humans and machines, etc.

说明：这是一个典型的长句，主句是"It is used in diverse areas"，其后的部分是宾语补足语。

译文：它应用于不同领域，如演示工程和科学计算及可视化结果，制作电视广告和专题片，模拟和分析现实世界问题，计算机辅助设计，增加人机间通信带宽的图形用户接口等。

[2] A structural method determines which pixels should be set before drawing the line, a conditional method tests certain conditions to find which pixel should be set next.

说明：这是一个并列句，每部分讲述一种方法，前部分的"which pixels should be set before drawing the line"是宾语从句；后部分的"to find which pixel should be set next"是宾语补足语。

译文：结构方法是在画线之前先决定像素的位置，条件方法是先验证一定的条件然后再决定像素的位置。

[3] A polygon, even though generally constructed from straight lines, is an important graphics primitive.

说明：本句中的"even though generally constructed from straight lines"是非限定性定语。

译文：多边形，尽管通常由直线组成，但它也是重要的图形元素。

[4] The characteristic stair step shape of raster lines is particularly noticeable on systems with low resolution, and we can improve their appearance somewhat by displaying them on high-resolution systems.

说明：本句是并列句，在前部分中，"The characteristic stair step shape"是主语，"raster lines"是定语；在后部分中，"we"是主语，"by displaying them on high-resolution systems"是方式状语。

译文：光栅线的这种特有的阶梯现象在低分辨率的系统上特别明显，我们可以通过使用高分辨率显示系统来改善这一点。

3.1.2　Text B

A Three-Dimensional (3-D) Graphics

Three-dimensional (3-D) graphics give realistic qualities to objects in computer programs, particularly computer games. 3-D graphics appear to have height, width, and depth. Although you view computer games on a two-dimensional (2-D) computer screen, modern technology creates a 3-D experience by adding the appearance of depth. A game programmer can give single objects or an entire virtual world a 3-D appearance.

Creating a 3-D appearance first requires that you create a wireframe. A wireframe is a series of lines, curves, and shapes arranged to resemble an object in a 3-D world. Most 3-D wireframes, for example, consist of a series of polygons. A completed wireframe enables you to identify the shape of the object, although it appears to hollow. To transform the appearance of the 3-D object from hollow to solid, you add a surface to the wireframe. Some 3-D graphics are composed of more than one wireframe. When adding a surface, it is important to make the object look as realistic as possible by adding color, texture, and reflectance. Reflectance refers to the amount of light the object's surface reflects.

With the surface added to a wireframe, you next consider how the object will be lit from one or more lighting sources. Some people create 3-D graphics using a technique called ray-tracing. Ray-tracing involves drawing an imaginary path that rays of light follow as they leave their source and then land on an object. The light intensity will be greater on some portions of the object and less on other portions. In addition, the object also might cast a shadow once it is lit from a particular angle.

When creating a 3-D world, the next considerations are perspective and depth of field. Perspective refers to differences in how objects appear in relation to one another when they are close to you, versus farther away. Objects appearing close may seem to be spaced apart. As they move farther away from you, they become closer to one another. A technique for calculating which objects appear in front of or behind one another is called the Z-Buffer, named after the imaginary axis from the screen to the distant horizon. Depth of field gives the appearance that objects further from you are less focused than closer objects.

Anti-aliasing is the final technique in creating 3-D objects that appear in a 3-D world. Anti-aliasing makes curved and diagonal lines appear straight. When computers render graphics with curved and diagonal lines, they often appear jagged. Anti-aliasing inserts additional colored pixels that give the appearance of a smooth edge. All these techniques combined create a realistic 3-D graphics.

The complex native of 3-D technology requires more computing power in order to render a graphic in an acceptable period of time. For example, computer gamers often buy computers designed for gaming so that a lack of performance does not show their game. Gaming computers often have faster processors, several gigabytes of RAM, and one or more video cards containing at least 256MB of RAM. These video cards also might support DirectX, which is a programming interface that allows game programmers direct access to enhanced hardware features. For some computer games to work properly, they require a video card that supports a specific version of DirectX.

Although game programmers spend many hours creating 3-D graphics for programs such as computer games, the results are rewarding when a player faces an experience so realistic that it is difficult to differentiate between the game and reality.

Key Words

anti-aliasing	抗锯齿，图形保真
axis	轴，轴线
diagonal	斜线，对角线
hollow	挖空（某物），变空
jag	使成锯齿状，使成缺口
perspective	透视，视角，观点
reflectance	反射率，反射系数
rewarding	值得的，令人满意的
surface	表面，表层
texture	纹理，结构，质地
wireframe	线框，线架加工

3.1.3 Exercises

1. Translate the following phrases into English

(1)虚拟世界
(2)计算机游戏
(3)像素比
(4)电子束
(5)输出图元

2. Translate the following phrases into Chinese

(1)three-dimensional graph
(2)straight line
(3)light intensity
(4)video card
(5)computing power

3. Identify the following to be True or False according to the text

(1)The polygon can be defined by an ordered sequence of vertices, the corners of the polygon.

(2)A pixel is a point of light.

(3)Implementing a polygon as a graphics primitive is natural and helpful.

(4)A primitive is a graphics object that is essential for the creation or construction of simple images.

(5)Width of field gives the appearance that objects further from you are less focused than closer objects.

4. Reading Comprehension

(1) To transform the _____ of the 3-D object from hollow to solid, you add a surface to the wireframe.

a. shape
b. surface
c. curve
d. appearance

(2) Digital devices display a _____ segment by plotting discrete points between the two endpoints.

a. polygon
b. chart
c. straight line
d. curved-line

(3) Point plotting is accomplished by converting a single coordinate position furnished by an application program into appropriate operations for the _____ in use.

a. device
b. output device
c. input device
d. input/output device

(4) We can define _____ as an image which consists of a finite ordered set of straight boundaries called edges.

a. polygon
b. line
c. pixel
d. chart

3.2 Multimedia

3.2.1 Text A

We break the word multimedia into its component part, we get multi-meaning more than one, and media-meaning form of communication. Those types of media include text, audio sound, static graphics images, animation and full-motion video. Because of what hardware can and cannot do, it is often a trade-off between a certain number of static graphics images, audio sound, full-motion video and text. As you may guess, textual information takes the least amount of space to store.

1. Text

Whether or not they have used a computer, most people are familiar with text. Text is the basis for word processing programs and is still the fundamental information used in many multimedia programs.

In fact, many multimedia applications are based on the conversion of a book to a computerized form. This conversion gives the user immediate access to the text and lets him or her display pop-up windows, which give definitions of certain words. Multimedia applications also enable the user to instantly display information related to a certain topic that is being viewed. Most powerfully, the computerized form of a book allows the user to look up information quickly (without referring to the index or table of contents).

The Windows operating environment gives the user an almost infinite range of expressing text. [1] By displaying text in more than one format, the message a multimedia application is trying to portray can be made more understandable.

One type of application, which many people use every day, is the Windows Help Engine.[2] This application is a text-based information viewer that makes accessing information related to a certain topic easy.

2. Audio Sound

The integration of audio sound into a multimedia application can provide the user with information not possible through any other method of communication. Some types of information can't be conveyed effectively without using sound. It is nearly impossible, for example, to provide an accurate textual description of the bear of a heart or the sound of the ocean. Audio sound can also reinforce the user's understanding of information presented in another type of media. For example, a narration might describe what is being seen in an animation clip. This can enhance the understanding of what the application is all about and lead to better comprehension. Experts in learning have found that presenting information using more than one sense aids in later retention of the information. Most importantly, it can also make the information more interesting to the user.

Audio sound is available in several different formats. Today, maybe the most common type of audio is red book audio. This is the standard specification used to refer to consumer audio compact discs. It is an international standard and is officially known as IEC 908. This specification is called red book audio because of the color of the cover of the publication that describes its formats. Red book audio sound can also be used in multimedia applications, and it forms the basis of the highest quality sound available.

Another audio sound format is the Windows wave file, which can be played only on PCs running the Windows operating environment. A wave file contains the actual digital data used to play back the sound as well as a header that provides additional information about the resolution and playback rate. Wave files can store any type of sound that can be recorded by a microphone.

The final type of audio sound that may be used is known as the Musical Instrument Digital Interface, or MIDI for short. The MIDI format is actually a specification invented by musical instrument manufacturers. Rather than being a digitized form of the sound, the MIDI specification is actually a set of messages that describes what musical note is being played. The MIDI specification cannot store anything except in the form of musical notes. MIDI music can be created with a sequencer.

3. Static Graphics Images

When you imagine graphics images you probably think of "still" images-that is, images such as those in a photograph or drawing. There is no movement in these types of picture. Static graphics images are an important part of multimedia because humans are visually oriented.[3] Windows is also a visual environment. This makes displaying graphics images easier than it would be in a DOS-based environment.

Static graphics images have a number of formats and can be created in a number of different ways. Just as you can see an unlimited number of photographs or pictures, the types of static graphics images that you can include in a multimedia application are almost unlimited.

4. Animation

Animation refers to moving graphics images. The movement of somebody giving CPR makes it much easier to learn cardiopulmonary resuscitation, rather than just viewing a static picture. Just as a static graphics image is a powerful form of communication, such is the case with animation. Animation is especially useful for illustrating concepts that involve movement. Such concepts as playing a guitar or hitting a golf ball are difficult to illustrate using a single photograph, or even s series of photographs, and even more difficult to explain using text. Animation makes it easier to portray these aspects of your multimedia application.

5. Full-Motion Video

Full-motion video, such as the images portrayed in a television, can add even more to a multimedia application. Although full-motion video may sound like an ideal way to add a powerful message to a multimedia application, it is nowhere near the quality you would expert after watching television.[4] Full-motion video is still in its beginning stages on PCs, and it is limited in resolution and size. Even with advanced methods of data compression, full-motion video can suck up hard disk space faster than water falls when poured out of a bucket.

Key Words

bear	忍受，支撑
bucket	桶，水桶
cardiopulmonary	心肺的
clip	夹子，卡子，剪
content	内容，含量，满足
describe	描写，叙述
guitar	吉他
infinite	无限的，无穷的

instantly	即时，立刻
integration	集成，整合，一体化
microphone	话筒
multimedia	多媒体
narration	叙述，故事
officially	官方地，正式地
playback	播放，重放，放音
portray	描绘，描写，描述
pour	倾泻，涌出，倒
publication	出版，公布，发表
reinforce	增强，加固
retention	保留，保持
resuscitation	复苏，复兴，再兴
sequencer	定序器，音序器
suck	吸入，吸取，吮
textual	文本的，教科书的
understandable	可理解的，可同情的

Notes

[1] The Windows operating environment gives the user an almost infinite range of expressing text.

说明："an almost infinite range of expressing text"是宾语"the user"的补足语。

译文：Windows 操作环境为用户表示文本提供了无限的能力。

[2] One type of application, which many people use every day, is the Windows Help Engine.

说明：本句的"which many people use every day"是非限定性定语从句。

译文：Windows 帮助引擎是许多人天天用到的一种多媒体应用程序。

[3] Static graphics images are an important part of multimedia because humans are visually oriented.

说明：本句由"because"引导原因状语从句。

译文：静态图像是多媒体的重要部分，因为人类是视觉定位的。

[4] Although full-motion video may sound like an ideal way to add a powerful message to a multimedia application, it is nowhere near the quality you would expert after watching television.

说明：本句由"although"引导让步状语从句。

译文：虽然全运动影像听起来像是一个往多媒体程序中加入强有力信息的理想方法，但它无法达到像人们看电视一样的效果。

3.2.2　Text B

Dreamweaver

Macromedia Dreamweaver is a professional HTML editor for visually designing and managing web sites and pages. Dreamweaver's visual editing features also let you quickly add design and functionality to your pages without writing a line of code. You can view all your site elements or assets and drag them from an easy-to-use panel directly into a document. Streamline your development workflow by creating and editing images in Macromedia Fireworks, then importing them directly into Dreamweaver, or by adding Flash objects you create directly in Dreamweaver.

Dreamweaver can display a document in three ways: in design view, in code view, and in a split view that shows both the design and code. To change the view in which you are working, select a view in the Dreamweaver toolbar. By default, Dreamweaver displays the document window in design view.

In addition, you can work with Dreamweaver's design view in two different ways in layout view and standard view. You select these views in the view category of the objects panel. In layout view you can design a page layout, insert graphics, text, and other media; in standard view, in addition to inserting graphics, text and media, you can also insert layers, create frame documents, create tables, and apply other changes to your page-options that are not available in layout view.

Adobe Photoshop

Adobe Photoshop is a kind of photo retouching, image editing, and color painting software. Whether you are a novice or an expert in image editing, the Photoshop program offers you the tools you need to get professional-quality results.

Photoshop provides integrated tools for creating and outputting crisp, editable vector shapes and text. With the new tools, you can incorporate resolution-independent, vector-based graphics and type into pixel-based images to achieve an unparalleled range of design effects.

The new rectangle, rounded rectangle, ellipse, polygon, and line tools let you create a wide variety of vector-based shapes. These tools can be used to create shape layers. Like Adobe Illustrator, Photoshop provides pathfinder operations—Add, Subtract, Restrict, and Invert—for quickly combining basic vector shapes into complex shapes.

Photoshop presents an intuitive new layer effects interface, a new selection of effect options, and new support for saving your layer effect designs as layer styles for ongoing use. The new Layer Styles dialog box shows at a glance which effects are applied to the currently selected layer and lets you define which effects to use in a layer style. Once you save a layer style, it appears in the new styles palette.

Authorware

Authorware is the multimedia making software that is put out by Macromedia Company. Authorware adopts the object-oriented programming and it is a kind of multimedia developing instrument on the basis of icon and process line. It gives numerous multimedia materials to other software and mainly undertakes integrating and organization work of multimedia materials itself. Because Authorware is a simple and powerful hypermedia creation tools, the range of application is very extensive. It has already applied in school teaching, enterprise training, various kinds of demonstrations of the reports, commercial field, and so on.

Key Words

crisp	卷曲，易碎的，脆的
ellipse	椭圆
functionality	功能性，功能
glance	一瞥，闪光，浏览
intuitive	知觉的，直观的
invert	颠倒，倒置
panel	面，板
pathfinder	探路者，探险者，开创者
professional	专业的，职业的
rectangle	长方形，矩形
retouch	润色，修饰
split	分开，拆分，分裂
unparallel	不平行的
workflow	工作流程

3.2.3　Exercises

1. Translate the following phrases into English

(1)代码视图
(2)数据压缩
(3)国际标准
(4)矢量图形
(5)文字的信息

2. Translate the following phrases into Chinese

(1)Design view
(2)media element
(3)playback rate
(4)full-motion video

(5)multimedia application

3. Identify the following to be True or False according to the text

(1)Authorware has already applied in school teaching, enterprise training, various kinds of demonstrations of the reports, commercial field, and so on.

(2)Macromedia Dreamweaver is a professional HTML editor for visually designing and managing web sites and pages.

(3)Photoshop does not provide integrated tools for creating and outputting crisp, editable vector shapes and text.

(4)Dreamweaver includes many coding tools and features.

(5)The types of media include text, audio sound, static graphics images, animation and full-motion video.

4. Reading Comprehension

(1)_____is the basis for word processing programs and is still the fundamental information used in many multimedia programs.

    a. Graphics image

    b. Sound

    c. Text

    d. Table

(2)Static graphics images have a number of _____and can be created in a number of different ways.

    a. data

    b. texts

    c. elements

    d. formats

(3)The integration of audio sound into a_____can provide the user with information not possible through any other method of communication.

    a. multimedia application

    b. text

    c. graphic image

    d. application

(4)By displaying text in more than one format, the message a multimedia application is trying to portray can be made more_____

    a. difficult.

    b. understandable.

    c. easier.

    d. quickly.

3.3 Computer Animation

3.3.1 Text A

Some typical applications of computer-generated animation are entertainment (motion pictures and cartoons), advertising, scientific and engineering studies, and training and education. Although we tend to think of animation as implying object motions, the term computer animation generally refers to any time sequence of visual changes in a scene.[1] In addition to changing object position with translations or rotations, a computer-generated animation could display time variations in object size, color, transparency, or surface texture.

Many applications of computer animation require realistic displays. An accurate representation of the shape of a thunderstorm or other natural phenomena described with a numerical model is important for evaluating the reliability of the model also, simulators for training aircraft pilots and heavy-equipment operators must produce reasonably accurate representations of the environment. Entertainment and advertising applications on the other hand, are sometimes more interested in visual effects. Thus, scenes may be displayed with exaggerated shapes and unrealistic motions and transformations. There are many entertainment and advertising applications that do require accurate representations for computer-generated scenes. And in some scientific and engineering studies, realism is not a goal. For example, physical quantities are often displayed with pseudo-colors or abstract shapes that change over time to help the researcher understand the nature of the physical process.

In general, a sequence is designed with the following steps:
- Storyboard layout
- Object definitions
- Key-frame specifications
- Generation of in-between frames

This standard approach for animated cartoons is applied to other animation applications as well, although there are many special applications that do not follow this sequence. Real-time computer animations produced by flight simulators, for instance, display motion sequences in response to settings on the aircraft controls.[2] And visualization applications are generated by the solutions of the numerical models. For frame-by-frame animations, each frame of the scene is separately generated and stored. Later, the frame can be recorded on film or they can be consecutively displayed in "real-time playback" mode.

The storyboard is an outline of the action. It defines the motion sequence as a set of basic events that are to take place. Depending on the type of animation to be produced, the storyboard could consist of a set of rough sketches or it could be a list of the basic ideas for the motion.[3]

An object definition is given for each participant in the action. Object can be defined in terms of basic shapes, such as polygons or splines. In addition, the associated movements for each object are specified along with the shape.

A key frame is a detailed drawing of the scene at a certain time in the animation sequence. Within each key frame, each object is positioned according to the time for that frame. Some key frames are chosen at extreme positions in the action; others are spaced so that the time interval between key frames is not too great. More key frames are specified for intricate motions than for simple, slowly varying motions.

In-betweens are the intermediate frames between the key frames. The number of in-betweens needed is determined by the media to be used to display the animation. Film requires 24 frames per second, and graphics terminals are refreshed at the rate of 30 to 60 frames per second. Typically, time intervals for the motion are set up so that there are from three to five in-betweens for each pair of key frames. Depending on the speed specified for the motion, some key frames can be duplicated. For a 1-minute film sequence with no duplication, we would need 1440 frames. With five in-betweens for each pair of key frames, we would need 288 key frames. If the motion is not too complicated, we could space the key frames a little farther apart.[4]

There are several other tasks that may be required, depending on the application. They include motion verification, editing, and production and synchronization of a soundtrack. Many of the functions needed to produce general animations are now computer-generated.

Some steps in the development of an animation sequence are well suited to computer solution, these include object manipulations and rendering, camera motions, and the generation of in-betweens. Animation packages, such as Wave-front, for example, provide special functions for designing the animation and processing individual objects.

One function available in animation packages is provided to store and manage the object database. Object shapes and associated parameters are stored and updated in the database. Other object functions include those for motion generation and those for object rending. Motions can be generated according to specified constraints using two-dimensional or three-dimensional transformations. Standard functions can then be applied to identify visible surfaces and apply the rendering algorithms.

Another typical function simulates camera movements. Standard motions are zooming, panning, and tilting. Finally, given specification for the key frames, the in-betweens can be automatically generated.

Key Words

consecutively	连续地
duplicate	复制品，副本
entertainment	娱乐，游艺
imply	暗示，意味着
interval	间隔，区间
intricate	错综的，复杂的

motion	运动，动机，意向
outline	大纲，提纲
panning	全景拍摄，移镜头
phenomenon（复数为 phenomena）	现象，事件
realistic	现实的，逼真的
render	表现，描写，反映
rotation	旋转，轮转
sketch	素描，草图，速写
soundtrack	音带，声迹
storyboard	剧本，故事情节
thunderstorm	雷雨
tilting	倾斜，倾卸

Notes

[1] Although we tend to think of animation as implying object motions, the term computer animation generally refers to any time sequence of visual changes in a scene.

说明：本句由 "although" 引导让步状语从句。

译文：尽管我们在考虑动画时倾向于想到暗指对象的移动，但术语 "计算机动画" 通常指场景中任何随时间而发生的视觉变化。

[2] Real-time computer animations produced by flight simulators, for instance, display motion sequences in response to settings on the aircraft controls.

说明：本句的 "produced by flight simulators" 是定语，修饰主语，"for instance" 是插入语，"to settings on the aircraft controls" 是目的状语。

译文：例如，飞行模拟器生成的实时计算机动画按飞机控制器上的动作来显示动画序列。

[3] Depending on the type of animation to be produced, the storyboard could consist of a set of rough sketches or it could be a list of the basic ideas for the motion.

说明：本句的 "Depending on the type of animation to be produced" 是分词结构作为状语，其后是并列句。

译文：依赖于要生成的动画类型，剧本可能包含一组粗略的草图或运动的一系列基本思路。

[4] If the motion is not too complicated, we could space the key frames a little farther apart.

说明：本句由 "if" 引导条件状语从句，"a little farther apart" 是宾语补足语。

译文：如果运动并不是很复杂，我们可以将关键帧安排得稀一点。

3.3.2　Text B

A large portion of digital image processing activity has been devoted to image restoration in history. Image restoration means the removal or reduction of degradation that were incurred while the digital image was being obtained. These degradations include the blurring that can be introduced by optical systems, image motion, and so on. The aim of image restoration is to bring the images toward what it would have been if it had been recorded without degradation. Degradation comes in many forms such as motion blur, noise, and camera mis-focus. In cases like motion blur, it is possible to come up with a very good estimate of the actual blurring function and "undo" the blur to restore the original image. In cases where the image is corrupted by noise, the best we may hope to do is to compensate for the degradation it caused.

The field of image restoration began primarily with the efforts of scientists involved in the space programs of both the United States and the former Soviet Union in the 1950s and early 1960s. These programs were responsible for producing many incredible images of the Earth and our solar system that, at that time, were unimaginable. Such images held untold scientific benefits which only became clear in the ensuing years as the race for the moon began to consume more and more of our scientific efforts and budgets. However, the images obtained from the various planetary missions of the time, such as the Ranger, Lunar Orbiter, and Mariner missions, were subject to many photographic degradations. These were a result of substandard imaging environments, the vibration in machinery and the spinning and tumbling of the spacecraft. Pictures from the later manned space missions were also blurred due to the inability of the astronaut to steady himself in a gravitationless environment while taking photographs. The degradation of images was no small problem, considering the enormous expense required to obtain such pictures in the first place. The loss of information due to image degradation could be devastating. For example, the 22 pictures produced during the Mariner IV flight to Mars in 1964 were later estimated to cost almost $10 million just in terms of the number of bits transmitted alone. Any degradation reduced the scientific value of these images considerably and clearly cost the space agencies money.

This was probably the first instance in the engineering community where the extreme need for the ability to retrieve meaningful information from degraded images was encountered. As a result, it was not long before some of the most common algorithms from one-dimensional signal processing and estimation theory found their way into the realm of what is today known as "digital image restoration." Digital image restoration is a very broad field, and thus contains many other successful approaches that have been developed from different perspectives, such as optics, astronomy, and medical imaging, and so on.

Digital image restoration is being used in many other applications as well. Just to name a few, restoration has been used to restore blurry X-ray images of aircraft wings to improve federal aviation inspection procedures. It is used for restoring the motion induced effects present in still composite frames (produced by the superposition of two temporally spaced fields of a video image), and, more generally, for restoring uniformly blurred television pictures. Printing applications often require the use of restoration to ensure that halftone reproductions of continuous images are of high quality. In addition, restoration can improve the quality of continuous images generated from halftone images. Digital restoration is also used to restore images of electronic piece parts taken in assembly-line manufacturing environments. Many defense-oriented applications require restoration, such as that of guided missiles, which may obtain distorted images due to the effects of pressure differences around a camera mounted on the missile. All in all, it is clear that there is a very real and important place for image restoration technology today.

Digital image restoration is a field of engineering that studies methods used to recover an original scene from degraded observations. It is an area that has been explored extensively in the signal processing, astronomical, and optics communities for some time. Many of the algorithms used in this area have their roots in well-developed areas of mathematics, such as estimation theory, the solution of ill-posed (inverse) problems, linear algebra and numerical analysis. Techniques used for image restoration are oriented toward modeling the degradations, usually blur and noise, and applying an inverse procedure to obtain an approximation of the original scene.

Image restoration is distinct from image enhancement techniques, which are designed to manipulate an image in order to produce results more pleasing to an observer, without making use of any particular degradation models. Image reconstruction techniques are also generally treated separately from restoration techniques, since they operate on a set of image projections and not on a full image. Restoration and reconstruction techniques do share the same objective, however, which is that of recovering the original image, and they end up solving the same mathematical problem, which is that of finding a solution to a set of linear or nonlinear equations.

Key Words

astronaut	航天员，宇航员
aviation	航空，飞机制造业
blurring	模糊，斑点甚多，混乱
budget	预算，预算案，预算拨款，按预算来计划
compensate	补偿，抵消
corrupt	堕落，腐败；错误百出的
degradation	退化，失真，降级
devastate	破坏，毁灭
enhancement	增强，提高，改善
federal	联邦的，同盟的
gravitationless	失重，无重力，无地心引力
halftone	半音，中间色，网板
meaningful	有意义的，有意图的
missile	导弹，投射物
mission	使命，派遣
planetary	行星的

projection	投影，投射，预测，规划
realm	领域，范围，界，境界
reconstruction	重建，再现
restoration	恢复，复原
spacecraft	宇宙飞船，航天器
spinning	使……旋转
superposition	重叠，叠加，重合
temporally	暂时地
tumble	倒塌，翻滚
unimaginable	难以想象的，想不到的

3.3.3 Exercises

1. Translate the following phrases into English

(1)时间间隔

(2)图像恢复

(3)实时回放

(4)伪彩色

(5)图像重建技术

2. Translate the following phrases into Chinese

(1)optical system

(2)in-between frame

(3)linear algebra

(4)image enhancement technique

(5)animation sequence

3. Identify the following to be True or False according to the text

(1)There are many entertainment and advertising applications that do require accurate representations for computer-generated scenes.

(2)The storyboard is not an outline of the action.

(3)More key frames are specified for intricate motions than for simple, slowly varying motions.

(4)Image restoration is distinct from image enhancement techniques.

(5)The field of image restoration began primarily with the efforts of scientists involved in the space programs.

4. Reading Comprehension

(1)For a 1-minute film sequence with no duplication, we would need_____frames.

  a. 640

  b. 720

c. 2880

d. 1440

(2) Film requires _____ per second, and graphics terminals are refreshed at the rate of 30 to 60 frames per second.

a. 42 frames

b. 36 frames

c. 24 frames

d. 20 frames

(3) Digital restoration is also used to restore _____ of electronic piece parts taken in assembly-line manufacturing environments.

a. images

b. words

c. papers

d. layers

(4) Using_____, it is possible to stack elements of movie clip one upon another orderly.

a. tool bar

b. animation sequence

c. time-axis window

d. layers

3.4 Virtual Reality

3.4.1 Text A

Virtual reality (VR) is a new technology, which emerged after the 1980s. But within several years, it had pervaded into various domains—science, technology, engineering, medicine, culture, entertainment, and its potential in application is just conspicuous.

With present achievements, using computer hardware and software and advanced sensors, researchers can generate a 3-dimensional artificial virtual environment, where you can walk around, watch in every direction and touch every object in the environment.[1] Everything in the environment is harmoniously combined with other objects so realistically that you may feel you are in a physical environment, but really you are roaming in a virtual world.

The newly uprising VR technology has given rise to innovation in all domains. For example, people can visit and examine a building or a plane before their blueprints come out. Medical students can be trained in a VR environment so that they would not hurt the patient and the risk of operation can be reduced to minimum.[2]

Virtual reality can be subdivided in many different ways. VR systems can be classified into three categories upon the visual channel.

♦ Head-mounted displays/ BOOMs

Head-mounted displays (HMDs), which typically also include earphones for the auditory channel as well as devices for measuring the position and orientation of the user, have been the primary VR visual device for much of the 1990s. Using CRT or LCD technology, HMDs provide two imaging screens, one for each eye. Thus, given sufficient computer power, stereographic images are generated. Typically, the user is completely immersed in the scene, although HMDs for augmented reality overlay the computer-generated image onto the view of the real world.[3]

An alternative to HMDs is the BOOM (Binocular Omni-Orientation Monitor). Two high-resolution CRTs are mounted inside a package against which the user places his eyes. By counterbalancing the CRT packaging on a free-standing platform, the display unit allows the user six-degree-of-freedom movement while placing no weight on the user's head.

HMDs and BOOMs are similar devices in that the user is fully immersed in the virtual environment and does not see his/her actual surroundings. The BOOM solves several of the limitations of the HMD (e.g., resolution, weight, field of view), but at the expense of reducing the sense of immersion by requiring the user to stand or sit in a fixed position.

♦ Immersive rooms

Immersion does not necessarily require the use of the head-mounted displays that are the most common method for presenting the visual channel in a virtual environment. The CAVE^{TM} (Cave Automatic Virtual Environment), a type of immersive room facility developed at the University of Illinois, Chicago, accomplishes immersion by projecting on two or three walls and a floor and allowing the user to interactively explore a virtual environment. An immersive room is typically about 10 foot×10 foot×13 foot (height), allowing a half-dozen or more users to examine the virtual world being generated within the space.

While HMDs require that users interact in virtual spaces (they can not see each other in their "real" environment), the immersive room offers the significant advantage of permitting user interaction, discussion, and analysis in the real world. However, the computational cost of generating scenes within an immersive room is very high. Two images must be generated at high refresh rates for each wall in the immersive room. In addition, each wall requires a high-quality projector, and since back projection is used, a large allocation of space is required for projection length. Costing over one-half million dollars, immersive rooms exist only in a handful of large research organizations and corporations.

♦ The VR Responsive Workbench

The VR Responsive Workbench operates by projecting a computer-generated, stereoscopic image off a mirror and then onto a table surface that is viewed by a group of users around the table. Using stereoscopic shuttered glasses, users observe a 3-D image displayed above the tabletop. By tracking the group leader's head and hand movements using magnetic sensors, the Workbench permits changing the view angle and interacting with the 3-D scene.[4] Other group members observe the scene as manipulated by the group leader, facilitating easy communication between observers about the scene and defining future actions by the group leader. Interaction is performed using speech recognition, a pinch glove for gesture recognition, and simulated laser pointer.

Key Words

achievement	完成，成就
auditory	听觉的，耳朵的
augment	扩张，扩大
binocular	双目，双筒望远镜
blueprint	蓝图，设计图
conspicuous	显著的，明显的
corporation	公司，有限公司
counterbalance	使平衡，弥补，抗衡，抵消
facilitate	便利，促进，使容易
gesture	姿势，表示
glove	手套
handful	一把，少数，少量
harmoniously	和谐地，相称地
hurt	损害，伤害
immerse	沉浸，浸入，陷入
immersion	沉浸，浸泡
limitation	限制，局限性
magnetic	有磁性的，磁性
mirror	镜子，反映
mount	安装，登上，架置
newly	新近，最近
orientation	方向，取向，定位
overlay	覆盖，重叠
pervade	扩大，普及
pinch	捏，掐，匮乏
projector	投影机
realistically	现实地，逼真地
roam	漫游，游历
sensor	传感器，感应器
significant	重要的，有意义的
stereographic	立体画法的，立体摄影术的
subdivide	再分，细分
surrounding	环境，周围

tabletop	桌面，台面
uprising	升起，立起
workbench	工作台，作业台

Notes

[1] With present achievements, using computer hardware and software and advanced sensors, researchers can generate a 3-dimensional artificial virtual environment, where you can walk around, watch in every direction and touch every object in the environment.

说明：这是一个长句，主句是"researchers can generate a 3-dimensional artificial virtual environment"，主句前的"With present achievements, using computer hardware and software and advanced sensors"是状语，由"where"引导非限定定语从句，修饰宾语"environment"。

译文：借助现有的成果，利用计算机硬件和软件及先进的传感器，研究人员能够创造三维人工虚拟环境，你可在其中漫步，四下观望和触摸环境中的每个物体。

[2] Medical students can be trained in a VR environment so that they would not hurt the patient and the risk of operation can be reduced to minimum.

本句由"so that"引导目的状语从句。

译文：医科学生能在 VR 环境中进行培训，这样就不会伤害病人，并能最大限度地降低手术的风险。

[3] Typically, the user is completely immersed in the scene, although HMDs for augmented reality overlay the computer-generated image onto the view of the real world.

本句由"although"引导让步状语从句。

译文：一般，虽然增加了逼真性的 HMD 把计算机生成的图像叠加在实际世界的视图上，但用户完全沉浸在此情景中了。

[4] By tracking the group leader's head and hand movements using magnetic sensors, the Workbench permits changing the view angle and interacting with the 3-D scene.

本句的"By tracking the..."是方式状语。

译文：通过使用磁传感器跟踪小组长之头与手的移动，工作台允许改变视角并且与三维场景进行交互。

3.4.2　Text B

Automatic speech recognition (ASR) is useful as a multimedia browsing tool: it allows us to easily search and index recorded audio and video data. Speech recognition is also useful as a form of input. It is especially useful when someone's hands or eyes are busy. It allows people working in active environment such as hospitals to use computers. It also allows people with handicaps such as blindness or palsy to use computers. Finally, although everyone knows how to talk, not as many people know how to type. With speech recognition, typing would no longer be a necessary skill for using a computer. If we ever were successful enough to be able to combine it with natural language understanding, it would make computers accessible to people who don't want to learn the technical details of using them.

In 1994, IBM was the first company to commercialize a dictation system based on speech recognition. Speech recognition has since been integrated in many applications: Telephony applications; Embedded systems; Multimedia applications, like language Learning Tools.

Many improvements have been realized since 60 years but computers are still not able to understand every single word pronounced by everyone. Speech recognition is still a very cumbersome problem.

There are quite a lot of difficulties. The main one is that two speakers, uttering the same word, will say it very differently from each other. This problem is known as inter-speaker variation. In addition the same person does not pronounce the same word identically on different occasions. This is known as intra-speaker variation. It means that even consecutive utterances of the same word by the same speaker will be different. Again, a human would not be confused by this, but a computer might. The waveform of a speech signal also depends on the recording conditions. Noise and channel distortions are very difficult to handle, especially when there is no a priori knowledge of the noise or the distortion.

A speech recognition system can be used in many different modes (speaker-dependent or independent, isolated / continuous speech, for small medium or large vocabulary).

A speaker-dependent system is a system that must be trained on a specific speaker in order to recognize accurately what has been said. To train a system, the speaker is asked to record predefined words or sentences that will be analyzed and whose analysis results will be stored. This mode is mainly used in dictation systems where a single speaker is using the speech recognition system. On the contrary, speaker-independent system can be used by any speaker without any training procedure. Those systems are thus used in applications where it is not possible to have a training stage (telephony applications, typically). It is also clear that the accuracy for the speaker-dependent mode is better compared to that of the speaker-independent mode.

Isolated word recognition is the simplest speech recognition mode and the less greedy in terms of CPU requirement. Each word is surrounded by a silence so that word boundaries are well known. The system does not need to find the beginning and the end of each word in a sentence. The word is compared to a list of words models, and the model with the highest score is retained by the system.

Continuous speech recognition is much more natural and user-friendly. It assumes the computer is able to recognize a sequence of words in a sentence. But this mode requires much more CPU and memory, and the recognition accuracy is really inferior compared with the preceding mode. Why is continuous speech recognition more difficult than isolated word recognition?

Some possible explanations are: speakers pronunciation is less careful; speaking rate is less constant; word boundaries are not necessarily clear; there is more variation in stress and intonation (interaction between vocal tract and excitation); additional variability is introduced by the unconstrained sentence structure; co-articulation is increased both within and between words; speech is mixed with hesitations, partial repetitions, etc.

Keyword spotting has been created to cover the gap between continuous and isolated speech recognition. Recognition systems based on keyword spotting are able to identify in a sentence a word or a group of words corresponding to a particular command.

The size of the available vocabulary is another key point in speech recognition applications. It is clear that the larger the vocabulary is the more opportunities the system will have to make some errors. A good speech recognition system will therefore make it possible to adapt its vocabulary to the task it is currently assigned to.

Key Words

accessible	可理解的，可达到的
blindness	失明，盲目
commercialize	商业化的，使商品化
cumbersome	累赘的，麻烦的，笨重的
dictation	听写，口授，命令
embedded	嵌入的，植入的，内含的
greedy	贪心的，贪婪的，渴望的
handicap	妨碍，缺陷，不利条件
hesitation	犹豫，踌躇
inferior	低劣的，下等的，较差的
palsy	瘫痪，中风
priori	先验的，先验，先天
pronounce	发音，读，宣称，演讲
recognition	识别，认可
utter	彻底的，完全的，绝对的
utterance	表达，说话，发声，话语
variation	变异，变化，变动
waveform	波形

3.4.3 Exercises

1. Translate the following phrases into English

(1)自动语音识别
(2)句子结构
(3)头戴式显示器
(4)虚拟空间
(5)三维虚拟模型

2. Translate the following phrases into Chinese

(1)dictation system

(2)virtual reality

(3)professional training

(4)virtual training device

(5)virtual military training

3. Identify the following to be True or False according to the text

(1)The size of the available vocabulary is not a key point in speech recognition applications.

(2)Isolated word recognition is the simplest speech recognition mode.

(3)The newly uprising VR technology has given rise to innovation in all domains.

(4)A speaker-dependent system is a system that must be trained on a specific speaker in order to recognize accurately what has been said.

(5)Immersion does necessarily require the use of the head-mounted displays.

4. Reading Comprehension

(1)An immersive room is typically about_____, allowing a half-dozen or more users to examine the virtual world being generated within the space.

a. 10 foot×12 foot×13 foot (height)

b. 10 foot×10 foot×10 foot (height)

c. 10 foot×13 foot×10 foot (height)

d. 10 foot×10 foot×13 foot (height)

(2)Virtual reality is a new technology, which emerged after the_____

a. 1960s

b. 1970s

c. 1980s.

d. 1990s

(3)In_____, IBM was the first company to commercialize a dictation system based on speech recognition.

a. 1994

b. 1996

c. 1998

d. 2000

(4)HMDs and_____are similar devices in that the user is fully immersed in the virtual environment and does not see his/her actual surroundings.

a. BOOMs

b. Immersive rooms

c. VR Responsive Workbench

d. VRML

3.5 Computer Aided Design

3.5.1 Text A

In the broadest sense, Computer Aided Design (CAD) refers to any application of a computer to the solution of design problems, the engineer may communicate with the computer in many forms, either via the visual display screen, keyboard, graph plotter or many more man-machine interfaces. They can ask a question and receive an answer from the computer in a matter of seconds. More specifically, CAD is a technique in which the engineer and a computer work together as a team, utilizing the best characteristics of each.[1]

CAD should involve the development of a central design description on which all applications in design and manufacture should feed. This implies that computer-based techniques for the analysis and simulation of the design, and for the generation of manufacturing instructions, should be closely integrated with the techniques for modeling the form and structure of the design. In addition, a central design description forms an excellent basis for the concurrent development of all aspects of a design in simultaneous engineering activities. In principle, CAD could be applied throughout the design process, but in practice its impact on the early stages, where very imprecise representation such as sketches are used extensively, has been limited. It must also be stressed that at present CAD does not help the designer in the more creative parts of design, such as the generation of possible design solutions, or in those aspects that involve complex reasoning about the design. For example in assessing by visual examination of drawings whether a component may be made, or whether it matches the specification. These aspects are, however, the subjects of considerable current research.

The computer is capable of holding vast quantities of information on permanent media such as magnetic disc or temporarily in immediate access store. It is therefore possible to represent the details of an engineering drawing or the shape of a car body in digital form and store this digital information in memory. This data can then be retrieved from memory, rapidly converted and displayed on a VDU graphics screen, or alternatively, plotted onto paper using a graph plotter. Besides, the designer can quickly and easily update or amend any part of the drawing. The drawing data can then be written back to memory in its updated form.

The function of CAD can be grouped into four different categories:
1. Design and geometric modeling
2. Engineering analysis
3. Kinematics
4. Drafting

In design and geometric modeling the engineer describes the basic shape of the component he or she is intending to make, and the computer converts these inputs into a mathematical model that it stores for later use.[2] Once the model has been created, the engineering analysis performed by the computer determines such fundamental parameters as weight, volume, structural strength, heating behavior, electrical conductivity, and so forth. With computer-simulated kinematics the user can determine whether or not the moving parts or other structures will interfere with the motion of the component being designed.[3] Finally, drafting routines are used to provide drawings and other pictorial representations for the manufacture of the desired component.

The general CAD system was developed by considering a wide range of possible uses of such a system. The following were considered in detail:

- Mechanical engineering design
- Building design
- Structural engineering design
- Electronic circuit design
- Animation and graphic design

It was thought that for most practical applications the general drawing system would be incorporated in a much larger specific applications system. For this reason the drawing system was as simple as possible consistent with reasonable running efficiency, so that it could be incorporated into an application system with the minimum of effort.[4]

For both the production of drawing items by analysis and the analysis of drawings, it is essential that there is a simple efficient link between data produced by the drawing system and analysis programs. It is also essential that graphic data can be annotated in a way which is recognized by analysis programs but which does not affect the drawing system.

Key Words

amend	修改，改善
annotate	注释，注解
assess	评定，评估
concurrent	同时发生的，并行的
conductivity	传导性，电导率，传导率
extensively	广泛地
imprecise	不精确的，不严密的
interfere	干预，干涉
kinematics	运动学
mathematical	数学的，数学模型
permanent	永久的，永恒的
pictorial	绘图的，形象的
plotter	绘图仪
principle	原则，原理
simultaneous	同时的，同时发生的

temporarily 暂时地，临时地

Notes

[1] More specifically, CAD is a technique in which the engineer and a computer work together as a team, utilizing the best characteristics of each.

说明："in which…"是一个介词前置的限定性定语从句，修饰和限定"a technique"。

译文：更确切地说，CAD 是使工程技术人员和计算机协同工作、彼此发挥长处的技术。

[2] In design and geometric modeling the engineer describes the basic shape of the component he or she is intending to make, and the computer converts these inputs into a mathematical model that it stores for later use.

说明：本句中的"In design and geometric modeling"是状语，由"and"引导两个并列句子，主语分别是"the engineer"和"the computer"。

译文：在设计和几何造型阶段，工程师描述将要制造的零件的基本形状，计算机把这些输入信息转换成数学模型，并将它存储起来供以后使用。

[3] With computer-simulated kinematics the user can determine whether or not the moving parts or other structures will interfere with the motion of the component being designed.

说明：本句的"With computer-simulated kinematics"是状语，"whether or not…"是宾语从句。

译文：借助计算机模拟运动学分析，设计人员可以确定运动部件或其他结构是否会干扰正在设计的零件的运动。

[4] For this reason the drawing system was as simple as possible consistent with reasonable running efficiency, so that it could be incorporated into an application system with the minimum of effort.

说明：本句的"so that"引导的是结果状语从句。

译文：由于此原因绘图系统应尽可能简单，而且运行效率高，因而这种结合不需要花很多精力。

3.5.2　Text B

Computer-aided manufacturing (CAM) is the technology concerned with the use of computer systems to plan, manage and control manufacturing operations through either direct or indirect computer interface with the plant's production resources. One of the most mature areas of CAM is numerical control, or NC. This is the technique of using programmed instructions to control a machine tool that grinds, cuts, mills, punches, bends, or turns raw stock into a finished part. The computer can now generate a considerable amount of NC instructions based on geometric data from the CAD database plus additional information supplied by the operator. Research efforts are concentrating on minimizing operator interactions.

Another significant CAM function is the programming of robots, which may operate in a workcell arrangement, selecting and positioning tools and workpieces for NC machines. These robots may perform individual tasks such as welding or assembly or carry equipment or parts around the shop floor.

Process planning is also a target of computer automation. The process plan may determine the detailed sequence of production steps required to fabricate an assembly from start to finish as it moves from workstation to workstation on the shop floor. Even though completely automatic process planning is almost impossible, but a process plan for a part can be generated if the process plans for similar parts already exist. For this purpose, group technology has been developed to organize similar parts into a family. Parts are classified as similar if they have common manufacturing features such as slots, chamfers, holes. Therefore, to automatically detect similarity among parts, the CAD database must contain information about such features. This task is accomplished by using feature-based modeling or feature recognition.

The analysis subprocess of the design is the area where the computer reveals its value. In fact, there are many available software packages for stress analysis, interference checking, and kinematic analysis. These software packages are classified as CAE. One problem with using them is the provision of the analysis model. It would not be a problem at all if the analysis model were derived automatically from the conceptual design. However, the analysis model is not the same as the conceptual design but is derived by eliminating unnecessary details from the design or by reducing its dimensions. The proper level of abstraction differs, depending on the type of analysis and the desired accuracy of the solution. Thus it is difficult to automate this abstraction process; accordingly the analysis model is often created separately. It is a common practice to create the abstract shape of the design redundantly by using a computer-aided drafting system or a geometric modeling system or sometimes by using the built-in capability of the analysis packages.

The analysis subprocess can be imbedded in the optimization iteration to yield the optimal design. Various algorithms for finding the optimal solution have been developed, and many optimization procedures are commercially available. Optimization procedures could be thought of as a component of CAD software, but it is more natural to treat optimization procedures separately.

The design evaluation phase also can be facilitated by use of the computer. If we need a design prototype for the design evaluation, we can construct a prototype of the given design by using software packages that automatically generate the program that drives the rapid prototyping machine. These packages are classified as CAM software. Of course, the shape of the prototype to be made should exist in advance in a type of data. The data corresponding to the shape are created by geometric modeling. Even though the prototype can be constructed conveniently with rapid prototyping, it would be even better if we could use a virtual prototype, often called digital mock-up, which provides the same valuable information.

Many software tools are also available for design optimization. Although design optimization tools may be regarded as CAE tools, they are commonly classified separately. Several research activities are under way to determine design shape automatically by integrating design optimization and analysis. The beauty of design analysis and optimization is that it allows the engineer to see how the product will behave and enables the engineer to catch any errors before going to the time and expense of building and testing physical prototypes. Because the cost of engineering goes up exponentially in the later stages of product development and production, the early optimization and refinement afforded by CAE analysis pays off in greatly reduced overall product development time and costs.

Key Words

bend	弯曲，拉弯，使成形
chamfer	倒角
geometric	几何学的
grind	磨，碾
iteration	迭代，重复，反复操作，循环
kinematic	运动学的
mill	铣，磨
prototype	原型，雏形，蓝本
punch	打孔，冲孔，冲床
redundantly	多余地，冗余地
refinement	精细，精致，优雅，得体
reveal	展现，启示，显露
welding	焊接法，焊接

3.5.3 Exercises

1. Translate the following phrases into English

(1)人机界面
(2)工程制图
(3)计算机辅助制造
(4)设计原型
(5)计算机辅助设计

2. Translate the following phrases into Chinese

(1)digital form
(2)manufacturing instruction
(3)building design
(4)electronic circuit design
(5)design evaluation phase

3. Identify the following to be True or False according to the text

(1)One of the most mature areas of CAM is numerical control.
(2)CAD should therefore enable the designer to tackle a task more quickly and accurately.

(3) At present CAD helps the designer in the more creative parts of design.
(4) The computer is capable of holding vast quantities of information on permanent media.
(5) The analysis subprocess can not be imbedded in the optimization iteration.

4. Reading Comprehension

(1) _____ refers to any application of a computer to the solution of design problems.

  a. Computer Aided Design
  b. Communication
  c. Artificial Intelligence
  d. Multimedia

(2) The process plan may determine the detailed _____ of production steps required to fabricate an assembly from start to finish as it moves from workstation to workstation on the shop floor.

  a. sequence
  b. plan
  c. form
  d. design

(3) At any point the designer can command the _____ to simulate the real-world behavior of the part being created.

  a. CAD
  b. interface
  c. computer
  d. display

(4) Computer aided design is so effective because the computer communicate with the _____ not with numbers but pictures.

  a. program
  b. computer
  c. display
  d. designer

3.6 专业英语中被动语态与长句的运用

3.6.1 被动语态的运用

专业英语文体在很多情况下是对某个科学论题的讨论，介绍某个科技产品和科学技术，为了表示一种公允性和客观性，往往在句子结构上采用被动语态描述，即以被描述者为主体，或者以第三者的身份介绍文章要点和内容。于是，被动语态反映了专业英语文体中文体的客观性。除表述作者自己的看法、观点外，被动语态很少直接采用第一人称表述法，但在阅读理解和翻译时，根据具体情况，又可以将一个被动语态句子翻译成主动形式，以便强调某个重点，更适合汉语的习惯。

被动语态在科技文章中用得十分频繁，主要有两个原因。第一个原因是科技文章重在描写行为或状态本身，所以由谁或由什么作为行为或状态的主体就显得不重要。行为或状态的主体或者没有必要指出，或者根本指不出来。被动语态使用频繁的第二个原因是便于向后扩展句子，避免句子头重脚轻和不平衡。

主动语态表示句子的主语是谓语动作的发出者；被动语态表示句子的主语是谓语动作的承受者。也就是说，主动语态句子中的宾语在被动语态中是句子的主语。由于被动语态句子的主语是谓语动作的承受者，故只有及物动词才会有被动语态。在科技英语中，为了着重说明客观事物和过程，被动语态用得更为广泛。凡是在不必、不知道或不愿说出主动者的情况下均可使用被动语态。

被动语态构成如下：

主语 + be +（及物动词）过去分词

1．常用被动语态的几种情况

（1）当我们强调的是动作的承受者或给动作的承受者较大关注时，多用被动语态。这时，由于动作的执行者处于次要地位，句子中 by 引导的短语可以省略。

例：The virus in the computer has been found out.

译文：计算机中的病毒已经找出来了。

（2）当我们不知道或不想说出动作的执行者时，可使用被动语态。这时，句子中不带由 by 引导的短语。

例：Electricity was discovered a very long time ago.

译文：电是在很久以前发现的。

（3）当动作的执行者是"物"而不是"人"时，常用被动语态。

例：Video can be recorded and transmitted in various physical media.

译文：视频文件可以在多种物理媒质中存储和传输。

（4）当动作的执行者已为大家所熟知，而没有必要说出来时，也常常使用被动语态。

例：Once this version has been tested and perhaps evaluated by the future user, more features are added and tested in an incremental manner until the system is complete.

译文：一旦这个版本经过测试，并且也许经过了未来用户的评估，更多的特性就可以逐渐地添加进去并且进行测试，直到完成系统。

（5）使用被动语态能更好地安排句子。

例：The professor came into the hall and was warmly applauded by the audience.

译文：教授走进大厅，大家热烈鼓掌。

2．科技英语中主要时态的被动语态形式

（1）一般现在时

一般现在时的被动语态构成如下：

主语 + am（is，are）+及物动词的过去分词

例：A database is designed, built, and populated with data for a specific purpose.

译文：数据库是由用于某种特定目的的数据设计、构造和提供的。

本句的谓语动词有三个，即"designed, built, and populated"，用被动语态表示客观性。

例：The information is preprocessed at the sender before it is transmitted over the communication channel and post-processed at the receiver.

译文：信息在通信渠道中传播之前由发送者进行了预处理，并且接收者收到信息后也要进行处理。

本句采用被动语态，主语是"The information"，谓语是"preprocessed"和"post-processed"。

例：The switches are used for the opening and closing of electrical circuits.

译文：开关是用来开启和关闭电路的。

（2）一般过去时

一般过去时的被动语态构成如下：

主语 + was（were）+及物动词的过去分词

例：He was made to finish repairing the printer.

译文：他被迫马上修好打印机。

例：He was asked to do the experiment at once.

译文：有人请他马上做实验。

（3）一般将来时

一般将来时的被动语态构成如下：

主语 + will be + 及物动词的过去分词

当主语是第一人称时，可用：

主语 + shall be + 及物动词的过去分词

例：I shall not be allowed to do it.

译文：不会让我做这件事的。

例：What tools will be needed for the job?

译文：工作中需要什么工具？

（4）现在进行时

现在进行时的被动语态构成如下：

主语 + is（are）being + 及物动词的过去分词

例：Electron tubes are found in various old products and are still being used in the circuit of some new products.

译文：在各种老产品里看到的电子管，在一些新产品的电路中也还在使用。

例：Our printer is being repaired by John.

译文：约翰正在修理我们的打印机。

（5）过去进行时

过去进行时的被动语态构成如下：

主语 + was（were）being + 及物动词的过去分词

例：The accident was being investigated.

译文：事故正在调查中。

例：The laboratory building was being built then.
译文：当时，实验大楼正在建造中。

（6）现在完成时

现在完成时的被动语态构成如下：

主语 ＋have（has）been ＋ 及物动词的过去分词

例：Solar batteries have been used in satellites to produce electricity.

译文：人造卫星上已经用太阳能电池发电。

例：New techniques have been developed by the research department.

译文：研究部门开发了新技术。

（7）过去完成时

过去完成时的被动语态构成如下：

主语 ＋had been ＋ 及物动词的过去分词

例：Electricity had been discovered for more than one thousand years by the time it came into practical use.

译文：电在发现一千多年之后，才得到实际应用。

例：When he came back, the problem had already been solved.

译文：他回来时，问题已经解决了。

3．含被动语态句子的翻译

在汉语中，也有被动语态，通常通过"把"或"被"等词体现出来，但它的使用范围远远小于英语中被动语态的使用范围，因此英语中的被动语态在很多情况下都翻译成主动结构。

（1）英语原文中的主语在译文中仍作为主语。在采用此方法时，我们往往在译文中使用"加以""经过""用……来"等词来体现原文中的被动含义。

例：In other words mineral substances which are found on earth must be extracted by digging, boring holes, artifical explosions, or similar operations which make them available to us.

译文：换言之，矿物就是存在于地球上，但须经过挖掘、钻孔、人工爆破或类似作业才能获得的物质。

例：Nuclear power's danger to health, safety, and even life itself can be summed up in one word: radiation.

译文：核能对健康、安全，甚至对生命本身构成的危险可以用一个词——辐射来概括。

（2）将英语原文中的主语翻译为宾语，同时增补泛指性词语（大家、人们等）作为主语。

例：Television, it is often said, keeps one informed about current events, allows one to follow the latest developments in science and politics, and offers an endless series of programmes which are both instructive and entertaining.

译文：人们常说，电视使人了解时事，熟悉科学与政治领域的最新发展，并能源源不断地为观众提供各种既有教育意义又有趣的节目。

例：It could be argued that the radio performs this service as well, but on television everything is much more living, much more real.

译文：可能有人会指出，无线电广播同样也能做到这一点，但还是电视屏幕上的节目更生动、真实。

（3）将英语原文中的 by、in、for 等作为状语的介词短语翻译成译文的主语，在此情况下，英语原文中的主语一般被翻译成宾语。

例：And it is imagined by many that the operations of the common mind can be by no means compared with these processes, and that they have to be acquired by a sort of special training.

译文：许多人认为，普通人的思维活动根本无法与科学家的思维过程相比，而且认为这些思维过程必须经过某种专门的训练才能掌握。

例：A right kind of fuel is needed for an atomic reactor.

译文：原子反应堆需要一种合适的燃料。

（4）专业英语中的一些被动句也可以翻译成汉语的被动句，常用"被""由……""受到""遭""为……所""使"等表示。

例：Over the years, tools and technology themselves as a source of fundamental innovation have largely been ignored by historians and philosophers of science.

译文：工具和技术本身作为根本性创新的源泉多年来在很大程度上被科学史学家和科学思想家们忽视了。

例：The behaviour of a fluid flowing through a pipe is affected by a number of factors, including the viscosity of the fluid and the speed at which it is pumped.

译文：流体在管道中流动的情况，受到流体黏度、泵送速度等各种因素的影响。

3.6.2 长句的运用

由于科学的严谨性，专业英语中常常出现许多长句。长句主要是由于修饰语过多、并列成分多及语言结构层次多等因素造成的，如名词后面的定语短语或定语从句，以及动词后面或句首的介词短语或状语从句。这些修饰成分可以一个套一个地连用（包孕式结构），形成长句结构。显然，英语的一句话可以表达好几层意思，而汉语习惯用一个小句表达一层意思，一般几层意思要通过几个小句来表达。在专业文章中，长句往往是对技术的关键部分的叙述，翻译不恰当就会错误地表达整个段落甚至通篇文章的意思。在阅读及翻译专业文章时，遇到长句要克服畏惧心理，无论多么复杂的句子也都是由一些基本成分组成的。要弄清楚原文的句法结构，找出整个句子的中心内容及其各层意思，并分析几层意思之间的逻辑关系。

对于长句的翻译而言，一遍看不出句子的意思，就多看几遍，在这过程中步步深入。从弄清楚句型、句种、结构和各成分之间的关系，逐渐推进；然后，纵观全局，确切地把握句子所表达的意思。从全局到局部，分清句子的整体结构；再从局部到总体，深入把握句子的细节。

例：Only by studying such cases of human intelligence with all the details and by comparing the results of exact investigation with the solutions of AI (Artificial Intelligence) usually given in the elementary books on computer science can a computer engineer acquire a thorough understanding of theory and method in AI, develop intelligent computer programs that work in a human like way, and apply them to solving more complex and difficult problems that present computers can not.

译文：只有很详细地研究这些人类智能情况，并把实际调查的结果与基础计算机科学书上给出的人工智能解决方案相比较，计算机工程师才能彻底地了解人工智能的理论和方法，开发出具有人类智能的计算机程序，并将其用于解决目前计算机不能解决的更复杂和更难的问题。

本句是复合句，一主一从。主句有一个主语、三个并列谓语。句子以"Only+状语"开头，主句的主、谓语部分倒装。

例：Those functions that describe how a computer software reacts to mechanical design, called application functions, are more often important to the engineer in engineering applications such as analysis, calculation, and simulation in connection with mechanical design because he wishes to known how they will work in use to which he wants to apply them.

译文：描述计算机软件如何适应机械设计的功能（称为软件的应用功能）对从事机械设计中的分析、计算和模拟仿真工作的工程师来说是很重要的，因为他希望了解这些功能在其所从事的工作中将如何发挥作用。

本句共有六个谓语动词：Describe，react to，are，wishes to know，will work 和 wants to apply，其主语分别为 that（代 functions），computer software，functions，he（代 the engineer），they（代 functions）和 he。其中，functions 句前没有关联词，是主句，其余为从句。整个句子的含义分两层：第一层的含义是计算机软件的功能在机械设计中对工程师很重要；第二层含义是解释为什么计算机软件的功能在机械设计中对工程师很重要。

例：The firewall prevents attacks from the outside against the machines in the inside network by denying connection attempts from unauthorized parties located outside, in addition, a firewall may also be utilized to prevent users behind the firewall from using certain services that are outside.

译文：防火墙通过拒绝来自外部的未授权方的连接企图，从而防止来自外部的对内部网络中的机器的攻击。除此之外，防火墙还可以用来阻止墙内用户使用外部服务。

这是一个长句，由"in addition"连接两个并列子句，"by denying connection attempts from unauthorized parties located outside"是方式状语，"behind the…"是定语。

通常分析长句时采用的方法如下：
- 找出全句的基本语法成分，即主语、谓语和宾语，从整体上把握句子的结构。
- 找出句子中所有的谓语结构、非谓语动词、介词短语和从句的引导词等。
- 分析从句和短语的功能，即是否为主语从句、宾语从句、表语从句等，若是状语从句，则分析它是属于时间状语从句、原因状语从句、条件状语从句、目的状语从句、地点状语从句、让步状语从句、方式状语从句、结果状语从句，还是属于比较状语从句。
- 分析词、短语和从句之间的相互关系，如定语从句修饰的先行词是哪一个等。
- 注意分析句子中是否有固定词组或固定搭配。
- 注意插入语等其他成分。

在英语长句的阅读和翻译过程中，必须清楚句子的逻辑结构、层次关系和所用的语体。

常用的翻译方法有以下几种：

◆ 顺序法

当英语长句的内容叙述层次与汉语基本一致时，或英语长句中所描述的一连串动作是按时间顺序安排的，可以按照英语原文的顺序翻译成汉语。

例：Being able to receive information from any one of a large number of separate places, carry out the necessary calculations and give the answer or order to one or more of the same number of places scattered around a plant in a minute or two, or even in a few seconds, computers are ideal for automatic control in process industry.

译文：由于计算机能从工厂大量分散的任何地方获取信息进行必要的运算，并在一两分钟甚至几秒钟内向分散在工厂各处的一处或多处提供响应或发出指令，所以它对加工工业的自动控制是非常理想的。

例：No such limitation is placed on an alternating-current machine; here the only requirement is relative motion, and since a stationary armature and a rotating field system have numerous advantages, this arrangement is standard practice for all synchronous machines rated above a few kilovolt amperes.

译文：交流电机不受这种限制，唯一的要求是相对运动，而且由于固定电枢和旋转磁场系统有很多优点，这种安排是所有额定电压超过几千伏安的同步电机的标准做法。

例：The close-loop system has a control unit which gets information from a sensing element, compares the real state with that required by the program and, when there is a different between the two, makes the necessary adjustment to the control element so that the desired state is maintained.

译文：闭环系统有一个控制单元，该单元从传感器获得信息，把真实值和程序的预定值进行比较，当两者有区别时，就对控制器做出必要的调整，从而保持预定值。

◆ 逆序法

逆序法指从长句的后面或中间译起，把长句的开头放在译文的结尾。这是由于英语和汉语的表达习惯不同：英语习惯采用前置性陈述，先结果后原因；而汉语习惯则相反，一般先原因后结果，层层递进，最后综合。当遇到这些表达次序与汉语表达习惯不同的长句时，就要采用逆序法。

例：Instead of paying someone to manually enter reams of data into the computer, you can use a scanner to automatically convert the same information to digital files using OCR (Optical Character Recognition) software.

译文：你只要在使用扫描仪的过程中借助于光学字符识别软件就可以将信息转换成数字文件的形式，从而代替人们手工将大量数据输入计算机中的过程。

例：In order to assist users to name files consistently, and, importantly, to allow the original creator and other users to find those files again, it is useful to establish naming conventions.

译文：为了帮助用户统一地命名文件，重要的是使最初的创建者和其他用户能再一次找到那些文件，建立命名公约是很必要。

◆ 分句法

有时，长句中的主语或主句与修饰词的关系并不十分密切，翻译时可以按照汉语多用短句的习惯，把长句的从句或短语化成句子，分开叙述。有时，英语长句包含多层意思，而汉语习惯用一个小句表达一层意思。为了使行文简洁，翻译时可把长句中的从句或介词短语分开叙述，顺序基本不变，保持前后的连贯。翻译时为了使语意连贯，有时需要适当增加词语。

例：The structure design itself includes two different tasks, the design of the structure, in which the sizes and locations of the main members are settled, and the analysis of this structure by mathematical or graphical methods or both, to work out how the loads pass through the structure with the particular members chosen.

译文：结构设计包括两项不同的任务：一是结构设计，确定主要构件的尺寸和位置；二是用数学方法或图解方法或二者兼用进行结构分析，以便在构件选定后计算出各载荷通过结构的情况。

例：The loads a structure is subjected to are divided into dead loads, which include the weights of all the parts of the structure, and live loads, which are due to the weights of people, movable equipment, etc.

译文：一个结构受到的载荷可以分为静载荷和动载荷两类。静载荷包括该结构各部分的重量。动载荷则是由于人和可移动设备等的重量而引起的载荷。

◆ 综合法

在一些长句单纯采用上述任何一种方法都不能准确翻译时，就需要仔细分析，或按照时间先后，或按照逻辑顺序，顺逆结合、主次分明地对全句进行综合处理。

例：Noise can be unpleasant to live even several miles from an aerodrome; is you think what it must be like to share the deck of a ship with several squadrons of jet aircraft, you will realize that a modern navy is a good place to study noise.

译文：噪声甚至会使住在距离飞机场几英里之处的人感到不适。如果你能想象到站在甲板上的几个中队喷气式飞机中间将是什么滋味的话，那你就会意识到现代海军是研究噪声的理想场所。

例：Modern scientific and technical books, especially textbooks, require revision at short intervals if their authors wish to keep pace with new ideas, observations and discoveries.

译文：对于现代科技书籍，特别是教科书，要是作者希望其内容与新见解、新观察、新发现保持一致，就应该每隔较短的时间就重新修改内容。

3.7 习题答案与参考译文

3.7.1 第 3 单元习题答案

3.1 Computer Graphics

1. Translate the following phrases into English

(1)虚拟世界　　　　　　　　　　virtual world

(2)计算机游戏　　　　　computer game
(3)像素比　　　　　　　pixel ratio
(4)电子束　　　　　　　electron beam
(5)输出图元　　　　　　output primitive

2. Translate the following phrases into Chinese

(1)three-dimensional graph　　三维图形
(2)straight line　　　　　　　直线
(3)light intensity　　　　　　光照强度
(4)video card　　　　　　　　显卡
(5)computing power　　　　　计算能力

3. Identify the following to be True or False according to the text

T　T　T　F　F

4. Reading Comprehension

(1)d　appearance
(2)c　straight line
(3)b　output device
(4)a　polygon

3.2　Multimedia

1. Translate the following phrases into English

(1)代码视图　　　　　　code view
(2)数据压缩　　　　　　data compression
(3)国际标准　　　　　　international standard
(4)矢量图形　　　　　　vector-based graphics
(5)文字的信息　　　　　textual information

2. Translate the following phrases into Chinese

(1)design view　　　　　设计视图
(2)media element　　　　媒体元素
(3)playback rate　　　　回放速度
(4)full-motion video　　全运动视频
(5)multimedia application　多媒体应用程序

3. Identify the following to be True or False according to the text

T　T　F　F　T

4. Reading Comprehension

(1)c　Text
(2)d　formats

(3) a   multimedia application
(4) b   understandable

3.3　Computer Animation

1. Translate the following phrases into English

(1) 时间间隔　　　　　　　　time interval
(2) 图像恢复　　　　　　　　image restoration
(3) 实时回放　　　　　　　　real-time playback
(4) 伪彩色　　　　　　　　　pseudo-color
(5) 图像重建技术　　　　　　image reconstruction technique

2. Translate the following phrases into Chinese

(1) optical system　　　　　　　光学系统
(2) in-between frame　　　　　　插值帧
(3) linear algebra　　　　　　　线性代数
(4) image enhancement technique　图像增强技术
(5) animation sequence　　　　　动画序列

3. Identify the following to be True or False according to the text

T　F　F　T　T

4. Reading Comprehension

(1) d　1440
(2) c　24 frames
(3) a　images
(4) d　layers

3.4　Virtual Reality

1. Translate the following phrases into English

(1) 自动语音识别　　　　　　automatic speech recognition
(2) 句子结构　　　　　　　　sentence structure
(3) 头戴式显示器　　　　　　head-mounted display
(4) 虚拟空间　　　　　　　　virtual space
(5) 三维虚拟模型　　　　　　3-dimensional virtual model

2. Translate the following phrases into Chinese

(1) dictation system　　　　　听写系统
(2) virtual reality　　　　　　虚拟现实
(3) professional training　　　专业训练
(4) virtual training device　　虚拟训练设备

(5) virtual military training　　　　虚拟军事训练

3. Identify the following to be True or False according to the text

F　T　T　T　F

4. Reading Comprehension

(1) d　10 foot×10 foot×13 foot (height)
(2) c　1980s.
(3) a　1994
(4) a　BOOMs

3.5　Computer Aided Design

1. Translate the following phrases into English

(1) 人机界面	man-machine interface
(2) 工程制图	engineering drawing
(3) 计算机辅助制造	computer-aided manufacturing
(4) 设计原型	design prototype
(5) 计算机辅助设计	Computer Aided Design

2. Translate the following phrases into Chinese

(1) digital form	数字形式
(2) manufacturing instruction	制造指令
(3) building design	建筑设计
(4) electronic circuit design	电路设计
(5) design evaluation phase	设计评估阶段

3. Identify the following to be True or False according to the text

T　T　F　T　F

4. Reading Comprehension

(1) a　Computer Aided Design
(2) a　sequence
(3) c　computer
(4) d　designer

3.7.2　第 3 单元参考译文

3.1　计算机图形学

3.1.1　课文 A

计算机图形学是计算机领域中一个了不起的发明。它应用于不同领域，如演示工程和科学计算及可视化结果，制作电视广告和专题片，模拟和分析现实世界问题，计算机辅助设计，增加人机间通信带宽的图形用户接口等。用计算机作图的技术已得到广泛应用，这对于探索计算机图形的本质是非常重要的。

图形的图元

图元是创造和组成复杂图像的基本图形对象。幸好图形是由三个基本图元构成的而不像图形的应用那样纷繁复杂。这些组件中最基本的元素是像素，即最小的图形元素。

一个像素是一个亮点。它在光栅扫描显示器上只是一个微小的点。尽管它没有结构，但它被定义为构成块，因此被认为是图形图元。CRT 的分辨率与点的尺寸（单个点的直径）有关。100 点每英寸对应的分辨率含义是一个点尺寸为 0.01 英寸。然而，实际上，像素是椭圆形而非圆形的。像素的形状仅取决于视觉显示单元的特性。两个相邻水平像素中心的距离与两个相邻垂直像素中心的距离之比称为像素比。

线，特别是直线，组成了计算机图像的重要构成块。例如，线是用于线形图、条形图和饼图，数学函数的二维、三维图像，工程制图和建筑设计的基本构成块。在计算机图形中，直线是创建图形的基础，我们称它为图形图元。画直线有两种不同的方法。结构方法是在画线之前先决定像素的位置，条件方法是先验证一定的条件然后再决定像素的位置。

多边形，尽管通常由直线组成，但它也是重要的图形元素。由于现实世界的大量的物体图示都是由多边形组成的，因此我们经常要将多边形作为一个独立的整体来对待。多边形是由直线或曲线围成的并用一种固定的颜色填充的封闭区域。因为图像是二维的，所以多边形是一个平面图形。

将多边形作为图形元素是很自然的且十分有益。我们可以把多边形定义为由一组有限有序的直线（边界）组成的图形。同样，多边形还可定义为一系列有序的顶点，即多边形的角。按一定的顺序横截顶点可以得到多边形的边。边界列表足以画出线框。两个相邻顶点定义一条边界。把最后一个顶点和第一个顶点相连就画出了这个多边形。

输出图元

通常，图形编程软件包提供了一些功能来描述场景，这些功能使用了称为输出图元的基本几何结构并将输出图元组合成更复杂的结构。每个输出图元是由输入的坐标数据和有关物体显示的一些信息来指定的。点和直线段是图的最简单的几何成分，其他可以用来构造图形的输出图元有：圆及其他圆锥曲线、二次曲面、样条曲线和曲面、多边形填色区域及字符串等。

通过将应用程序提供的单个坐标位置转换成输出设备的相应操作，可以进行点的绘制。例如，对于 CRT 监视器，则是打开电子束，从而在选中的位置上照亮屏幕的荧光层。电子束的定位方法取决于显示技术。

通过计算沿线路径上两个指定端点间的中心位置，可以绘制一条线段，输出设备则按指令在端点间的这些位置上直接填充。对于矢量笔绘图仪或随机扫描显示器这类模拟设备，可以从一个端点到另一个端点绘制光滑线段。这是根据 x 方向和 y 方向需要修改的实际量，线性地改变水平和垂直偏转电压而实现的。

数字设备通过绘制两个端点间的离散点来显示线段。线路路径上离散的坐标位置是通过直线方程计算出来的。对于光栅视频显示器，线段的颜色（亮度）放入对应于像素位置的帧缓冲器中。视频控制器从帧缓冲器中读入该亮度值，然后显示在屏幕的像素上。一般使用整数值表示屏幕位置。这样将坐标值取整使得显示的线段具有阶梯现象（锯齿形）。光栅线的这种特有的阶梯现象在低分辨率的系统上特别明显，我们可以通过使用高分辨率显示系统来改善这一点。更有效的平滑光栅线的技术，则是基于对沿线路径上像素强度的调整。

3.1.2 课文 B
三维（3-D）图形

三维（3-D）图形能够使计算机程序中，尤其是计算机游戏中的物体更具真实感。3-D图形是看上去具有高、宽、深的图形。虽然用户是在二维（2-D）计算机屏幕上观看计算机游戏，但现代技术能通过给外观添加深度来创造三维的体验。游戏程序员可以使单个物体或整个虚拟世界具有三维外观。

创建三维外观，首先要创建线框。线框是一系列布置好的直线、曲线和图形，看起来像三维世界中的物体。例如，大多数三维线框是由一系列多边形组成的。尽管它看上去是空心的，但完整的线框能使用户识别出物体的形状。要使三维物体的外观由空心变为实心，需要给线框添加表面。有一些三维图形是由一个以上的线框组成的。添加表面时，重要的是通过添加颜色、纹理和反射率使物体看上去尽可能逼真。反射率是指物体表面反射的光量。

给线框添加完表面后，接下来要考虑如何用一个或多个光源照亮物体。有些人使用称为光线跟踪的技术来创建三维图形。光线跟踪包括要绘制出光线从光源到物体所跟踪的假想路径。物体上一些区域的光照强度会大些，另一些区域的会小些。此外，当从某个特定的角度照亮物体时，物体也可能会投下影子。

创建三维世界时，下一个要考虑的是透视和景深。透视是指物体接近你与远离你相比，物体彼此之间是如何显示出不同的。看起来近的物体，也许相隔很远。当它们离你越来越远时，它们看起来彼此更接近。一种用来计算物体是出现在另一个物体的前面还是后面的技术称为Z缓冲。Z缓冲的名字来自从屏幕向外延伸的虚轴。景深使离你远的物体看上去比离你近的物体更模糊。

抗锯齿是创建出现在三维世界里的三维物体的最终技术。抗锯齿能够使曲线和斜线看上去很平滑。当计算机渲染带有曲线和斜线的图形时，它们经常出现锯齿。抗锯齿能够插入额外的彩色像素，让边缘看上去更平整。将所有这些技术结合起来能够创建出逼真的三维图形。

三维技术的复杂性使它需要更多的计算能力才能在可接受的时间内渲染图形。例如，计算机游戏玩家通常需要购买为游戏设计的计算机，这样才不会因性能不足而无法显示游戏。游戏计算机往往具有速度更快的处理器、数 GB 的内存，以及一个或多个至少 256MB 缓存的显卡。这些显卡还可能支持 DirectX——一个允许游戏程序员直接访问可增强硬件功能的可编程接口。要使一些计算机游戏可以正常工作，它们需要支持特定版本的 DirectX 的显卡。

尽管游戏程序员为计算机游戏之类的程序创建三维图形要花很多时间，但当玩家体验到如此逼真，以至难以区分游戏和现实时，这些努力就是值得的。

3.2 多媒体

3.2.1 课文 A

如果我们将 multimedia 这个词分开，我们便得到 multi——意为多于一个，和 media——意为通信的形式。媒体类型包括文本、声音、静态图像、动画和动态视频。基于硬件的能力，通常要权衡选择静态图像、声音、全运动影像和文本，可以推测，文本信息占用存储空间最小。

1．文本

无论是否用过计算机，大多数人都熟悉文本。文本是字处理程序的基础，并且仍作为许多多媒体程序中的基本信息。

实际上，许多多媒体应用程序基于由书本到计算机形式的转换，这种转换使用户能直接访问文本，并为其显示一个弹出式窗口，以给出特定词的定义。多媒体应用程序也使用户能够立刻显示与正在浏览的某个主题有关的信息。功能更强的是，一本计算机化的书允许用户快速查找信息（而不必根据索引或目录）。

Windows 操作环境为用户表示文本提供了无限的能力。通过以多种形式显示文本，使人们更易理解多媒体应用程序要表达的信息。

Windows 帮助引擎是许多人天天用到的一种多媒体应用程序。该应用程序是一种基于文本的查询界面，可方便地访问某一主题的相关信息。

2．声音

把声音融入多媒体程序，用户可以得到使用其他通信方式无法得到的信息。某些类型的信息不用声音很难有效表达，例如用文字准确描述心脏的跳动声及大海的声音几乎是不可能的。声音也可以加深用户对其他媒体表示的信息的理解，例如可以把看到的动画片讲述出来。它可以帮助人们理解应用程序从而更好地理解多媒体。学习研究专家已经发现用多种感官表达的信息对信息的后期记忆很有帮助。最重要的是，多媒体信息可以引起用户的更大兴趣。

声音有几种不同格式。今天，也许最普遍的声音类型是红皮书声音，这是用于消费者 CD 声音的标准说明，它作为一种国际标准并正式公布为 IEC 908。之所以被称为红皮书是由于描述这种格式的书的封面是红色的。红皮书声音也被用于多媒体程序，它是获得高质量声音的基础。

另一种声音格式是 Windows 声波文件，它只能用于 Windows 操作环境下的 PC。声波文件包含用于回放声音的实际数字数据和文件头，文件头提供有关分辨率及回放速度的附加信息。声波文件可以存储通过麦克风录入的所有声音。

最后一种声音格式称为乐器数字接口，缩写为 MIDI。MIDI 格式实际上是由乐器制造商制定的，MIDI 实际上并非数字化的声音，而是描述要演奏的音符的信息集合。MIDI 不能存储除音符外的任何东西。MIDI 音乐可以由定序器生成。

3．静态图像

当你想象图像时，你可能想到静态图像——也就是像照片或画一样，这种类型的图像是不动的。静态图像是多媒体的重要部分，因为人类是视觉定位的。Windows 也是可视化环境，它比基于 DOS 的环境更容易显示图像。

静态图像有许多种格式，也可以用许多不同的方式生成。就像你可以看到无数的照片或图像一样，多媒体应用程序中包括的静态图像类型几乎也是无限的。

4．动画

动画就是运动的图像，给出 CPR 运动图像比只给一幅静态图像使人更容易明白心肺复苏运动。动画和静态图像一样，都是强有力的通信形式，动画在解释涉及运动的概念方面特别有用。讲解如何弹吉他或打高尔夫球，只用一幅图是不行的，甚至用一系列的图也不行，用文字讲解就更困难了，而在多媒体应用程序中用动画讲解则轻而易举。

5．全运动影像

全运动影像，例如电视里的图像，可使多媒体的应用更为广泛。虽然全运动影像听起来像是一个往多媒体程序中加入强有力信息的理想方法，但它无法达到像人们看电视一样的效果。PC 上的全运动影像仍处于初始阶段，其大小和分辨率仍受限制。即使用高性能的数据压缩算法，全运动影像的数据也极快地占满硬盘空间，其速度比瀑布还快。

3.2.2 课文 B

Dreamweaver

Macromedia Dreamweaver 是为可视化设计并管理网站和网页的一个专业的 HTML 编辑器。Dreamweaver 的可视化编辑特征也让你快速地把设计和功能加到你的页面中，而不用写一行代码。你能查看所有的你的站点的元素或资源，并可把它们直接从一个容易使用的面板拖进一个文件中。通过在 Micromedia Fireworks 中创建并编辑图像来优化开发工作流程，然后将它们直接导入 Dreamweaver 中，或加入直接在 Dreamweaver 创造的 Flash 对象中。

Dreamweaver 能以三种方式显示一个文件：设计视图、代码视图、设计与代码结合视图。在 Dreamweaver 工具栏中选择视图，就可以改变正在工作的视图。Dreamweaver 的默认设置是以设计视图显示文档窗口。

此外，在 Dreamweaver 的设计视图下，你能工作于布局视图和标准视图两种不同的方式。可以在对象面板的视图分类中选择这些视图。在布局视图中可以设计页面布局，插入图形、文本和其他媒体元素。在标准视图下，除插入图形、文本和媒体外，还可以插入图层、创建帧文件、创建表，以及对页选项实现其他在布局视图中无法做到的改变。

Adobe Photoshop

Adobe Photoshop 是具有照片修正、图像编辑和彩色绘画功能的软件。不论你是一位初学者，还是一位图像编辑方面的专家，Photoshop 程序都为你提供了获得专业水准作品所需的多种工具。

为了更好地进行图像的制作与输出，Photoshop 提供了编辑矢量图形和文本的一套完整工具。用这些新的工具，你可以将与分辨率无关的矢量图形、字体和像素图像进行混合编辑，以获得无与伦比的设计效果。

新的矩形、圆角矩形、椭圆、多边形、直线等工具，可以用来生成各种矢量图形。这些工具可以用来制作图形层。像 Adobe Illustrator 一样，为快速地将基本矢量图形合并为复杂的图形，Photoshop 提供了并、交、差、补等寻径操作。

Photoshop 提供了一种新的直观的图层效果界面、新的效果选择，并可将你设计的图层效果作为图层样式存储，以供日后使用。新的图层样式对话框使你一眼就能看出对当前图层施用的效果，并且还允许你在图层样式中对所使用的效果进行定义。一旦你将其存储为图层样式，它就会出现在新样式模板中。

Authorware

Authorware 是由 Macromedia 公司推出的多媒体制作软件，Authorware 采用面向对象的程序设计，是一种基于图标（Icon）和流程线（Line）的多媒体开发工具。它把众多的多媒体素材交给其他软件处理，本身则主要承担多媒体素材的集成和组织工作。由于 Authorware 是一个简单易用且功能强大的超媒体创作工具，因此其应用范围十分广泛。目前，它已经应用于学校教学、企业培训、各种演示报告、商业领域等。

3.3　计算机动画

3.3.1　课文 A

计算机生成动画的代表性应用有娱乐（电影和卡通片）、广告、科学和工程研究，以及培训和教学。尽管我们在考虑动画时倾向于想到暗指对象的移动，但术语"计算机动画"通常指场景中任何随时间而发生的视觉变化。除通过平移、旋转来改变对象的位置外，计算机生成的动画还可以随时间进展而改变对象大小、颜色、透明性和表面纹理等。

许多计算机动画的应用要求有真实感的显示。利用数值模型来描述的雷暴雨或其他自然现象的精确表示对评价该模型的可靠性是很重要的。同样，培训飞机驾驶员和大型设备操作员的模拟器必须生成环境的精确表示。另一方面，娱乐和广告应用有时较为关心视觉效果。因此可能使用夸张的形体和非真实感的运动和变换来显示场景。但确实有许多娱乐和广告应用要求计算机生成场景的精确表示。在有些科学和工程研究中，真实感并不是一个目标。例如，物理量经常使用随时间而变化的伪彩色或抽象形体来显示，以帮助研究人员理解物理过程的本质。

通常，一个动画序列按照以下步骤进行设计：

◆ 故事情节拆分
◆ 对象定义
◆ 关键帧描述
◆ 插值帧的生成

这种制作动画片的标准方法也适用于其他动画应用，尽管有许多专门的应用并不按此序列进行处理。例如，飞行模拟器生成的实时计算机动画按飞机控制器上的动作来显示动画序列。而可视化应用则由数值模型的结果生成。对于逐帧动画，场景中的每一帧是单独生成和存储的。然后，这些帧可以记录在胶片上或以"实时回放"模式连续地显示出来。

剧本是动作的轮廓。它将动画序列定义为一组要发生的基本事件。依赖于要生成的动画类型，剧本可能包含一组粗略的草图或运动的一系列基本思路。

为动作的每个参加者给出对象定义。对象可能使用基本形状如多边形或样条曲线进行定义。另外，每个对象的相关运动则根据形体而指定。

一个关键帧是动画序列中特定时刻的一个场景的详细图示。在每个关键帧中，每个对象的位置依赖于该帧的时刻。选择某些关键帧作为行为的极端位置。另一些则以不太大的时间间隔进行安排。对于复杂的运动，要比简单的缓慢变化运动安排更多的关键帧。

插值帧是关键帧之间过渡的帧。插值帧的数量取决于用来显示动画的介质。电影胶片要求 24 帧/秒，而图形终端按 30~60 帧/秒来刷新。一般情况下，运动的时间间隔设定为每对关键帧之间有 3~5 个插值帧。依赖于为运动指定的速度，有些关键帧可重复使用。一分钟没有重复的电影胶片需要 1440 帧。如果每两个关键帧之间有 5 个插值帧，则需要 288 幅关键帧。如果运动并不是很复杂，我们可以将关键帧安排得稀一点。

可能还要求其他一些依赖于应用的任务。包括运动的验证、编辑和声音的生成与同步。生成一般动画的许多功能现在都由计算机来完成。

开发动画序列中的某几步工作很适合由计算机进行处理。其中包括对象管理和绘制、照相机运动和生成插值帧。动画软件包，如 Wave-front，提供了设计动画和处理单个对象的专门功能。

动画软件包中有存储和管理对象数据库的功能。对象形状及其参数存于数据库中并可更新。其他的对象功能包括运动的生成和对象绘制。运动可依赖指定的约束，使用二维或三维变换而生成。然后可使用标准函数来识别可见曲面并应用绘制算法。

另一种典型功能是模拟照相机的运动，标准的运动有拉镜头、摇镜头和倾斜。最后，给出对关键帧的描述，然后自动生成插值帧。

3.3.2 课文 B

数字图像处理的大部分研究一直都致力于图像恢复。图像恢复是指去除或减小所获数字图像的失真。这些失真可能是光学系统导致的模糊或图像抖动。图像恢复的目的是把图像恢复成它本该记录的无失真形式。失真表现为多种形式，如运动模糊、噪声、相机未聚焦等。出现运动模糊时，可以采用实际的模糊函数，以达到很好的估计，从而"消除"模糊现象，恢复原始图像。在图像含噪声时，我们希望最好能补偿由此造成的失真。

图像恢复领域的研究是在 20 世纪 50 年代和 60 年代初参与美国和苏联的太空计划的科学家们的努力下开始的。太空计划中获得了许多对地球和我们的太阳系来说难以置信的图像，这在当时是难以想象的。这些图像获得了无以言表的科学价值，使后来的登月竞赛更为激烈，使科学探索和预算大增。然而，当时利用测距仪、月球轨道器、水手号飞船等飞行任务获得的行星图像，大都存在因为摄影导致的失真。这是由于成像环境不标准、机械振动和航天器的旋转翻滚等产生的。后来的载人航天飞行得到的图像也是模糊的，这是由于航天员不能在一个稳定的环境中拍照。因为失重，他们在拍照的时候无法站稳。图像失真不是小问题，因为费用巨大，所以要在第一时间获得这些图像。由于图像失真造成的损失可能是灾难性的。例如，在 1964 年水手 IV 号飞船飞往火星期间提供的 22 幅图像，后来估计仅在数据传输方面就几乎花费了 1000 万美元。任何失真都会大大降低这些图像的科学价值，无疑会让太空机构付出更大的代价。

在极端情况下，鼓励利用失真的图像找回有意义的信息，这可能是工程领域遇到的第一个实例。因此，不久之后，一维信号处理和估计理论中的一些最常见的算法就进入了当今称为"数字图像恢复"的领域。数字图像恢复是一个非常宽广的领域，包含了光学、天文学、医学影像等各领域成功的图像恢复方法。

数字图像恢复也广泛应用于其他领域。仅举例如下：图像恢复已用于恢复飞机机翼模糊的 X 射线图像，以改进联邦航空检验过程。数字图像恢复用于去除合成帧中静止图像因运动造成的影响（由视频图像的两个时间间隔场叠加而成），以及用于恢复模糊的电视图像。印刷领域通常需要使用图像恢复技术，以确保对连续图像的中间色实现高品质的复制。此外，图像恢复可改善由中间色图像生成的连续图像的质量。数字图像的恢复也可用于恢复在电子装配线制造环境中取得的电子部件的图像。许多防御性领域也需要图像恢复，如导弹领域，由于导弹上照相机周围压力差的影响，可能使图像失真。所有的这些应用都清楚地表明，今天的图像恢复技术具有非常现实和重要的地位。

数字图像恢复是一个研究图像恢复方法的工程领域，利用这些方法可恢复受损的图像。它广泛应用于信号处理、天文、光学等领域。在这一领域中使用的许多算法基于大力发展的数学理论，如估计理论、病态逆问题求解、线性代数、数值分析等。图像恢复技术以建立劣化模型为目的，劣化模型包括模糊或含有噪声的图像，应用逆运算来获得接近原始场景的图像。

图像恢复不同于图像增强技术，图像增强的目的是为了产生令观察者满意的图像，而不需要使用任何特定的劣化模型。一般来说，图像重建技术也不同于恢复技术，前者是对一组图像的投影数据而不是全部图像进行处理。但恢复与重建技术有相同的目标，即恢复原来的图像，它们最终解决的是同一个数学问题，即找到一个线性或非线性方程组的解。

3.4 虚 拟 现 实

3.4.1 课文 A

虚拟现实（VR）是一项在 20 世纪 80 年代后出现的新技术。然而仅仅几年，就已渗透到各个领域——科学、技术、工程、医学、文化、娱乐，并且它的应用潜力的确引人注目。

借助现有的成果，利用计算机硬件和软件及先进的传感器，研究人员能够创造三维人工虚拟环境，你可在其中漫步，四下观望和触摸环境中的每个物体。这种环境中的一切事物均与其他物体如此真实和谐地结合在一起，以致你可能觉得自己处在一个真实的物理环境中，但实际上你却是在一个虚拟世界中漫游。

新兴的 VR 技术已经使各个领域的革新层出不穷。例如，人们有可能在绘制出一座建筑物或一架飞机的蓝图之前，就对其进行参观考察。医科学生能在 VR 环境中进行培训，这样就不会伤害病人，并能最大限度地降低手术的风险。

虚拟现实可按很多不同的方式来细分。基于视觉通道，VR 系统可分为三个范畴。

◆ 头戴显示器/BOOM

头戴显示器（HMD）一般包括听觉通道用的耳机，以及测量用户的位置与方向的一些设备，在 20 世纪 90 年代的大多数日子里，它们一直是主要的 VR 视觉设备。使用 CRT 或 LCD 技术，HMD 提供两个图像屏，每只眼一个。因此，如果计算机功能足够强的话，立体的图像就被生成。一般，虽然增加了逼真性的 HMD 把计算机生成的图像叠加在实际世界的视图上，但用户却完全沉浸在此情景中了。

HMD 的替代品是 BOOM（双筒全方位监控器），两个高分辨率 CRT 被安装在一个标准部件内，用户的眼睛就对着它。通过使安装在独立式平台上的 CRT 平衡后，显示部件允许用户在其头上未放置重量时有 6 个自由度的移动。

HMD 和 BOOM 是类似的设备，其类似性在于用户完全沉浸在虚拟环境中而不查看其实际的周围环境。BOOM 解决了 HMD 的几个局限性（如分辨率、重量、视野），但其代价是要求用户站或坐在固定的位置而减少了沉浸的感觉。

◆ 沉浸室

头戴式显示器是演示虚拟环境中视觉通道的最普遍的方法，但沉浸并不一定要求使用头戴式显示器。CAVE^{TM}（Cave 自动虚拟环境）是由芝加哥伊利诺斯大学开发的一类沉浸室设计，通过投影到两面或三面墙面及地板上，并允许用户交互地探测虚拟环境的办法来完成沉浸。一个沉浸室一般大约是 10 英尺×10 英尺×13 英尺（高度），允许 6 个或更多的用户检查在该空间内正被生成的虚拟世界。

当 HMD 需要用户在虚拟空间交互作用（他们不能在他们的"实际"环境中彼此看到）时，沉浸室提供的重要好处是允许用户在实际世界中进行交互、讨论和分析。然而，在沉浸室内生成情景的计算费用是十分高的，必须以高刷新率为沉浸室中的每面墙生成两个图像。另外，每面墙需要高质量的投影仪，并且因为使用了背投，为了投影长度需要配给大的空间。因为成本超过 50 万美元，所以沉浸室仅存在于少数大的研究机构和公司。

◆ VR 响应工作台

VR 响应工作台的操作方法是把一个计算机生成的立体图像投影到一面镜子上反射出来，然后投射到一个台子表面上，围绕该台子的一群用户观察该台子表面。使用有快门的立体双筒镜，用户观察显示在台面上方的一个三维图像。通过使用磁传感器跟踪小组长之头与手的移动，工作台允许改变视角并且与三维场景进行交互。小组其他成员随着小组长的操纵而观察场景，这样便于在观察者之间就场景进行交流，并由小组长确定进一步的一些行动。交互作用是使用语音识别、姿势识别专用连指手套及模拟的激光指示器来完成的。

3.4.2 课文 B

自动语音识别（ASR）是一种非常有用的多媒体浏览工具。它使我们可以轻易地查找和索引已记录的音频和视频数据。语音识别同样是一种很有用的输入方式，当人们的双手和眼睛忙碌时就更是如此。它使工作在医院等繁忙环境中的人们可以使用计算机。同样，它也可以使失明或瘫痪等残疾人使用计算机。毕竟，虽然每个人都知道如何说话，但却没有同样多的人知道如何打字。有了语音识别，打字不再是一项使用计算机所必需的技巧。一旦我们能足够成功地把语音识别和自然语言理解相结合，那么计算机就会被那些不想学习使用其技术细节的人们所接受。

在 1994 年，IBM 是第一个将基于语音识别的听写系统商业化的公司。自那时起，语音识别被集成到很多应用中：电话应用；嵌入式系统；多媒体应用，如语言学习工具。

60 年来，我们实现了很多改进，但计算机仍然不能够理解每个人所发出的每个单词。语音识别仍然是一个十分艰难的问题。

很多困难确实存在。一项主要困难是，当两个人发同一个单词的音时，每个人会发得非常不同。这个问题被称为不同人差异。除此之外，同一个人在不同的场合下发同一个单词的音是不同的，这被称为同人差异。这意味着即使是同一个人连续发出同一个单词也是不相同的。此外，人类不会对此感到困惑，但计算机却有可能。语音信号的波形依赖录音条件。对噪声和信道失真是很难处理的，特别是当没有对噪声和失真的先验知识时更是如此。

语音识别系统可以被用于许多不同模式（对发言者的依赖或独立，孤立或连续发言，小媒介或大词汇）下。

发言者依赖系统是一种必须针对特定发言者进行训练的系统，以便准确地识别他所说的内容。为训练该系统，发言者必须录制事先定义的将被分析的词和句子，分析结果将被存储起来。这种模式主要用于听写系统，该系统是由单一发言者使用的语音识别系统。相反，发言者独立系统可以在无训练的情况下被所有人使用。因此，这种系统被用于不可能有训练阶段的应用（典型的如电话应用）中。显而易见，发言者依赖模式的准确性要好于发言者独立模式。

孤立单词识别是一种最简单的语音识别模式，在对 CPU 的需求上也是最不贪婪的一种。每个单词都被寂静所包围，因此单词间的界限是很明显的。这种系统并不需要寻找句子中的每个单词的开始与结束，而是会将它们与一系列的单词模型进行比较，最高分的模型会被系统保留。

连续语音识别更加自然，对用户也更加友好。它假设计算机能够识别句子中的一系列单词。但这种模式需要更多的 CPU 和内存空间，而且识别的准确性也不如前述模式。为什么连续语音识别要比孤立单词识别更困难呢？

一些可能的原因是：发言者的发音不标准；发言速率不恒定；单词间的界限不够清晰；在重音和语调上会存在更多变异（方言和情绪之间相互作用）；由于句子结构不受约束也会产生其他变异；在单词内或单词之间连续现象增加；发言伴随着犹豫停顿，部分重复等。

关键字定位是用来填补连续和孤立语音识别之间的空白而被创造的。基于关键字定位的识别系统能够从句子中辨认具有特定指示含义的一个或一组单词。

可使用的词汇量是语音识别应用的又一个关键点。显然，词汇量越大，系统出错的可能性也越大。因此，好的语音识别系统可以根据当前承担的任务来调整自己的词汇量。

3.5 计算机辅助设计

3.5.1 课文 A

广义上，计算机辅助设计（CAD）是指利用计算机解决设计问题。工程技术人员可以借助可视化显示屏、键盘、绘图仪和更多的人机接口与计算机通信。工程技术人员可以提出问题并能很快从计算机得到解答。更确切地说，CAD 是使工程技术人员和计算机协同工作、彼此发挥长处的技术。

CAD 涉及核心设计描述的开发，它决定了所有的设计和制造。这意味着用于分析和模拟设计及产生制造指令的以计算机为基础的技术，应该与设计的形式和结构建模技术密切结合。另外，核心设计描述为同时存在的模拟工程活动设计的各方面的开发打下了良好的基础。原则上，CAD 能够应用于设计的全过程，但实际上，它在开始阶段广泛使用草图等非常不准确的描述，所造成的影响使它具有一定的局限性。还必须强调的是，目前 CAD 还不能帮助设计者设计更富有创造性的产品，比如生成可能的设计方案，或那些涉及设计复杂推理的方面。例如，通过目视检查图纸来评估是否可以制造组件，或者它是否符合规格。然而，这些方面是当前正在研究的课题。

计算机可以将大量的信息保存在磁盘这样的永久性介质上或临时存储在当前访问的存储器中。因此，以数字形式描述一个工程图纸的细目或一个汽车车身的造型，并把信息存储在存储器中都是可以做到的。这些数据能从存储器中检索、快速转换并显示在视频显示器图形屏幕上，或交替地利用绘图仪绘制在纸上。此外，设计者还可以迅速、方便地更新或修改图纸的任何部分。也能把修改后的图纸数据写回到存储器中。

计算机辅助设计的功能可以归结为 4 种不同的类型：
1. 设计和几何造型
2. 工程分析
3. 运动学分析
4. 作图

在设计和几何造型阶段，工程师描述将要制造的零件的基本形状，计算机把这些输入信息转换成数学模型，并将它存储起来供以后使用。数学模型建立后，由计算机完成的工程分析将确定重量、体积、结构强度、受热特性、导电性能等基本参数。借助计算机模拟运动学分析，设计人员可以确定运动部件或其他结构是否会干扰正在设计的零件的运动。最后，绘图程序将为制造商绘制所要求的部件的图纸及图形化的表达信息。

在研究通用 CAD 系统时要考虑该系统应具有尽可能广的应用范围，仔细考虑以下几个方面：
- 机械工程设计
- 建筑设计
- 结构工程设计
- 电子电路设计
- 动画和图形设计

对大多数实际应用来讲，应考虑把通用绘图系统结合到大型专用系统中，由于此原因绘图系统应尽可能简单，而且运行效率高，因而这种结合不需要花很多精力。

对分析绘图项和绘图分析这两个过程而言，在由绘图系统所产生的数据和分析程序之间必须有一简单有效的联系。另外，图形数据可用能被分析程序识别但不对绘图系统产生影响的方法进行注释。

3.5.2 课文 B

计算机辅助制造（CAM）是涉及用计算机系统进行计划、管理和通过计算机与工厂生产资源直接或间接接口以控制制造过程的技术。CAM 最成熟的领域之一就是数控技术，即 NC 技术。这是使用程序指令来控制机床进行磨、切、铣、冲、折弯或车削，将坯料加工成零件的技术。目前，计算机可用基于 CAD 数据库的几何数据加上由操作人员提供的其他信息生成大量的 NC 指令。人们正在努力研究使操作人员的干预降至最少。

CAM 的另一个重要功能就是为机器人编程，它可以按工作单元的安排，为数控设备选择定位工具及工件。这些机器人可以完成像焊接或组装这样的单项任务，也可在车间内搬运设备或零件。

　　工艺计划也是计算机自动化的一个目标，工艺计划可以决定在车间里从一个工作台到另一个工作台、从头到尾地完成组装的生产步骤的详细顺序。尽管工艺计划完全自动化几乎是不可能的，但一个零件的工艺计划可以从已存在的相似零件的工艺计划加以概括得到。为此目的，群组技术已被开发出来用于将相似零件归为一类。如果具有相似的制造特性，如开槽、倒角、打孔的零件就分为同一类。因此，为了自动检测零件的相似性，CAD 数据库必须包括此类特征信息。这个任务是由采用基于特征的建模或特征识别技术来实现的。

　　设计过程的分析子过程最能体现计算机的价值。实际上，对于应力分析、干扰检验和运动分析有许多现成的软件包。这些软件包归入 CAE 一类。使用这些软件包的一个问题是如何提供分析模型。如果能从概念设计中自动导出分析模型，则不成问题。但是，分析模型与概念设计不同，它是通过在设计中忽略了不必要的细节或缩小尺寸而得来的。根据分析模型和所期望求解精度的不同，就有不同的抽象水平。因此，很难对抽象过程进行自动处理，而分析模型通常是分别建立的。常用的做法是，用计算机辅助绘图系统或几何建模系统或有时使用分析软件包的内置功能，为设计过程多建立几个抽象模型。

　　分析子过程可以嵌入优化循环中以获取最优设计。获取最优解决方案的各种算法都已被开发出来，并且也可在市场上购买许多优化程序。优化过程可以认为是 CAD 软件的一个部分，但最好还是单独对待。

　　通过使用计算机也可促进设计评估阶段。如果为了设计评价而需要一个设计原型，则可以通过使用软件包来构建给定设计的原型，这些软件包可以自动生成驱动快速原型机的程序。这些软件包都归类为 CAM 软件。当然，要做的原型形状应事先变成某种类型的数据。对应于该形状的数据应该由几何建模系统来创建。尽管原型可以方便地用原型速成方法来构建，但如果使用虚拟原型就会更加理性。虚拟原型通常称为数字模型，它提供了同样有价值的信息。

　　对于设计优化也有许多软件工具。尽管可以认为设计优化工具是 CAE 工具，但通常它们不归为同一类。有一些正在研发的项目是通过集成设计优化和分析来自动确定设计形状的。设计分析和优化的妙处就在于它能让工程师了解产品的工作状态，并在花费时间、资金来构建和测试物理原型之前就发现所有错误。因为工程造价在产品开发和生产后期呈指数趋势增加，所以由 CAE 分析得来的早期优化和细化就得到了回报——大大减少了整个产品开发的时间和费用。

Unit 4　Computer Network and Its Applications

4.1　Computer Network Basics

4.1.1　Text A

Network establishes communication among computers. This system is especially helpful when people work on different place. It improves the speed and accuracy of communication, prevents messages from being misplaced and automatically ensures total distribution of key information.

A network consists of several computers linked by communication lines. The machines can perform the functions independently, but their activities can also be coordinated. Originally the aim was the exchange of information (program, data files) between the users of different mainframes. When smaller computers became available and spread in large numbers within a single organization, connections between these small computers became attractive for the shared use of resources (printer, disk, processing time) as well. The resulting systems, in which computers of possibly different type and size are connected but fully retain their autonomy, are called computer networks.

A local area networks, or LAN, is a communication network that is privately owned and that covers a limited geographic area such as an office, a building, or a group of building. Hardware resource sharing allows each personal computer in the network to access and use devices that would be too expensive to provide for each user. Information resource sharing allows anyone using a personal computer on the local area network to access data stored on any other computer in the network.[1] A wide area network, or WAN, is geographic in scope (as opposed to local) and uses telephone lines, microwaves, satellites, or a combination of communication channels.

The configuration, or physical layout, of the equipment in a communication network is called topology. Communication networks are usually configured in one or a combination of three patterns. These configurations are star, bus, and ring networks. Although these configurations can be used with wide area networks, we illustrate them with local area networks. Devices connected to a network, such as terminal, printers, or other computers, are referred to as nodes.

- ◆ Star Network

A star network contains a central computer and one or more terminals or personal computers connected to it, forming a star.[2] A pure star network consists of only point-to-point lines between the terminals and the computer, but most star networks include both point-to-point lines and multi-drop lines. A star network configuration is often used when the central computer contains all the data required to process the input from the terminals, Such as an airline reservation system. For example, if inquiries were being processed in the star network, all the data to answer the inquiry would be contained in the database stored on the central computer.

A star network can be relatively efficient, and close control can be kept over the data processed on the network. Its major disadvantage is that the entire network is dependent on the central computer and the associated hardware and software.[3] If any of these elements fail, the entire network is disabled. Therefore, in most large star networks, backup computer systems are available in case the primary system fails.

◆ Bus Network

When a bus network is used, all the devices in the network are connected to a single cable. Information is transmitted in either direction from any one personal computer to another. Any message can be directed to specific device. An advantage of the bus network is that devices can be attached or detached from the network at any point without disturbing the rest of the network. In addition, if one computer on the network fails, this does not affect the other users of the network.

◆ Ring Network

A ring network does not use a centralized host computer. Rather, a circle of computers communicate with one another. A ring network can be useful when the processing is not done at a central site, but at local sites. For example, computers could be located in three departments: accounting, personnel, and shipping and receiving. The computers in each of these departments could perform the processing required for each of the departments. On occasion, however, the computer in the shipping and receiving department could communicate with the computer in the accounting department to update certain data stored on the accounting department computer. Data travels around a ring network in one direction only and passes through each node. Thus, one disadvantage of a ring network is that if one node fails, the entire network fails because the data does not get past the failed node.[4] An advantage of a ring network is that less cable is usually needed and therefore network cabling costs are lower.

Key Words

airline	航线，航空公司
attach	附加，附属
attractive	有吸引力的
autonomy	自治，自治权，自主
backup	备份
centralize	集中，聚集
channel	通道，频道

detach	使分离，分遣
disadvantage	缺点
distribution	分配，分布
geographic	地理的
illustrate	说明，表明
inquiry	查询，探索
microwave	微波，微波炉
misplace	放错地方
node	网络节点，连接到网络上的设备
occasion	时机，场合
originally	起初，原来，最初
prevent	预防，阻碍，阻止
privately	私自，私下地
pure	纯正的，纯的
reservation	保留，预订
retain	保持，保留
ring	环，环形物，形成环形
satellite	卫星

Notes

[1] Information resource sharing allows anyone using a personal computer on the local area network to access data stored on any other computer in the network.

说明：本句中的"anyone"是宾语，"using a personal computer on the local area network"是"anyone"的定语，"to access data"是宾语补足语。

译文：信息资源共享允许局域网上每个计算机用户访问存储在网上其他计算机中的数据。

[2] A star network contains a central computer and one or more terminals or personal computers connected to it, forming a star.

说明：本句中的"a central computer and one or more terminals or personal computers"是宾语。

译文：星形网络由一台中央计算机和一台或多台连接到该中央计算机上并形成星形结构的终端或计算机组成。

[3] Its major disadvantage is that the entire network is dependent on the central computer and the associated hardware and software.

说明：本句由"that"引导的是表语从句。

译文：其主要缺点是整个网络都依赖于中央计算机及其相关的硬件和软件。

[4] Thus, one disadvantage of a ring network is that if one node fails, the entire network fails because the data does not get past the failed node.

说明：本句中的"that"引导表语从句，"if one node fails"是条件状语，而"because the

data does not get past the failed node"是原因状语。

译文：因此，环形网络的缺点是如果一个节点出现故障，由于数据不能通过出现故障的节点，就会使整个网络无法工作。

4.1.2 Text B

The Internet is an international web of interconnected government, education, and business computer networks—in essence, a network of networks. A person at a computer terminal or personal computer with the proper software communicates across the Internet by placing data in an Internet Protocol (IP) packet—an electronic envelope—and "addressing" the packet to a particular destination on the Internet. Communications software on the intervening networks between the source and destination networks "read" the addresses on packets moving through the Internet and forward the packets toward their destinations.

How To Use Internet

Windows provides you with quick and easy access to the Internet, no matter what programs you are running or tasks you are performing. The Active Desktop allows you to customize your workspace and the Address bar helps you to connect to the Internet from any window. You can also find a variety of tools that help you to communicate with people and other computers.

Most people connect to the Internet by using network connection or Internet service provider (ISP). An ISP supplies a service number that you can dial from your computer to log on the Internet server. Once you connect to the system, you have access to the Internet, e-mail, and any other services supplied by your ISP. Your ISP also furnishes you with the details you need to configure an Internet connection on your computer. If you use a network connection, your system administrator provides this information for you.

With the communication tools included in Windows, you can use your computer to send e-mail, handle phone calls, send a fax, or conduct a meeting with a video conference. For example, you can use Phone Dialer to answer phone calls or join a video conference in your company. With Internet Explorer and an Internet connection, you can search for and view information on the World Wide Web. You can type the address of the Web page that you want to visit into the address bar, or click an address from your list of favorites. Internet Explorer also lets you search the Internet for people, businesses, and information about subjects that interest you.

Internet Services

The most popular and widespread Internet application services include:
(1) World Wide Web (WWW)
WWW is a large network of Internet servers providing hypertext and other services to terminals running client applications such as a browser. WWW enables users to search, access, and download information from a worldwide series of networked servers where information is

dynamically interlinked.

(2) Electronic mail (E-mail)

Electronic mail allows a user to compose memos and send them to individuals or groups. Another part of the mail application allows users to read memos that they are received. There are two kinds of E-mail protocol used in the Internet. One is Simple Mail Transfer Protocol (SMTP) which accepts incoming connections and copies messages from them into the appropriate mailboxes. Another is Post Office Protocol-3 (POP3) which fetches E-mail from the remote mailbox and stores it on the user's local machine to be read later.

(3) File Transfer Protocol (FTP)

FTP is an application protocol, part of the TCP/IP protocol stack, used for transferring files between network nodes. The TCP/IP protocols include a file transfer application program that allows users to send or receive arbitrarily large files of programs or data.

(4) Telecommunication network (Telnet)

Telnet is used for remote terminal connection, enabling users to log into remote systems and use resources as is they were connected to a local system.

Key Words

administrator	管理员，管理人员
arbitrarily	任意地，擅自，肆意地
browser	浏览器，浏览程序
dial	拨号，拨号盘
dynamically	动态地
envelope	信封，封皮
favorite	特别喜爱的，中意的
furnish	布置，提供，陈设
interlink	连接，衔接
intervening	介入，干涉
mailbox	邮箱，信箱
memo	备注，备忘录
protocol	协议，议定书，草案
provider	供应者，提供者

4.1.3 Exercises

1. Translate the following phrases into English

(1)电子邮件
(2)简单邮件传输协议
(3)文件传输协议
(4)广域网

(5)环网

2. Translate the following phrases into Chinese

(1)Internet service provider

(2)star network

(3)backup computer system

(4)Local Area Network

(5)World Wide Web

3. Identify the following to be True or False according to the text

(1)A star network does not use a centralized host computer.

(2)In most large star networks, backup computer systems are available.

(3)Windows provides you with quick and easy access to the Internet, no matter what programs you are running or tasks you are performing.

(4)A ring network can transmit information from any one personal computer to another.

(5)WAN uses telephone lines, microwaves, satellites, or a combination of communication channels.

4. Reading Comprehension

(1)An _____ supplies a service number that you can dial from your computer to log on the Internet server.

 a. FTP

 b. ISP

 c. IP

 d. TCP

(2)With _____ and an Internet connection, you can search for and view information on the World Wide Web.____

 a. FTP

 b. network

 c. electronic mail

 d. Internet Explorer

(3)When a _____ is used, all the devices in the network are connected to a single cable.

 a. bus network

 b. ring network

 c. star network

 d. network

(4)Communications software on the intervening networks between the source and destination networks "read" the addresses on packets moving through the Internet and forward the packets toward their _____.

a. destinations
b. addresses
c. databases
d. sources

4.2 Wireless Network

4.2.1 Text A

In a wireless network, the computers are connected by radio signals instead of wires or cables. Advantages of wireless networks include mobility and no unsightly wires. Disadvantages can include a slower connection than a wired network and interference from other wireless devices. Your computer needs an internal or external wireless network adapter. To see if your computer has a wireless network adapter, do the following:

Open Network Connections by clicking the "Start" button, clicking "Control Panel", clicking "Network and Internet", clicking "Network and Sharing Center", and then clicking "Manage Network Connections". The adapters installed in your computer are listed.

In the list of available wireless networks, you will see a symbol that shows the wireless signal strength for each network. A strong signal usually means that the wireless network is close or there is no interference. For best performance, connect to the wireless network with the strongest signal. However, if an unsecured network has a stronger signal than a security-enabled one, it is safer for your data if you connect to the security-enabled network.[1] To improve the signal strength, you can move your computer closer to the wireless router or access point, or move the router or access point so it is not close to sources of interference such as brick walls or walls that contain metal support beams.[2]

Public wireless networks are convenient, but if they are not properly secured, connecting to one might be risky. Whenever possible, only connect to wireless networks that require a network security key or have some other form of security, such as a certificate. The information sent over such networks is encrypted, which can help protect your computer from unauthorized access.[3] In the list of available, each network is labeled as either security-enabled or unsecured. If you do connect to a network that is not secure, be aware that someone with the right tools can see everything that you do, including the websites you visit, the documents you work on, and the user names and passwords you use. You should not work on documents or visit websites that contain personal information, such as your bank records, while you are connected to that network.

Wireless vendors face the challenge of supporting increasingly bandwidth hungry applications, such as voice over IP, streaming video and videoconferencing. To dramatically increase throughput, 802.11a proponents had to solve a major challenge of indoor radio frequency. They had to develop a way to resolve the problem of delay spread in the current 2.4GHz, single-carrier and delay-spread system.

Delay spread is caused by the echoing of transmitted radio frequency. As these signals proceed to a certain point, such as a wireless antenna, they often bounce and echo off objects, walls, furniture and floors, and arrive at the antenna at different times due to the different path lengths.[4] A baseband processor, or equalizer, is required to "unravel" the divergent radio frequency signals. The delay spread must be less than the symbol rate, or the rate at which data is encoded for transmission. If not, some of the delayed signal spreads into the next symbol transmission. This can put a ceiling on the maximum bit rate that can be sustained.

With current bit-rate technology, this ceiling tends to be around 10M to 20M bit/sec. The 802.11a standard cleverly solves this challenge through an innovative modulation technique called Coded Orthogonal Frequency Division Multiplexing (COFDM), which has found earlier application in European digital TV and audio transmission. COFDM breaks the ceiling of the data bit rate by (1) sending data in a massively parallel fashion, and (2) slowing the symbol rate down so each symbol transmission is much longer than the typical delay spread. A guard interval (sometimes called a cyclic prefix) is inserted at the beginning of the symbol transmission to let all delayed signals "settle" before the baseband processor demodulates the data.

Wireless vendors now have a goal to boost wireless throughput beyond 100M bit/sec. While the 802.11a standard currently tops out at 54M bit/sec in 20MHz channels, several firms are developing and proposing high-rate extensions to the 802.11a standard. These proposals generally envision at least doubling throughput to anywhere from 108M to 155M bit/sec.

There is a newly development network for LANs: WLAN (Wireless LAN). Using electromagnetic waves, WLANs transmit and receive data over the air, minimizing the need for wired connections. Thus, WLANs combine data connectivity with user mobility, and through simplified configuration, enable movable LANs. Over the recent several years, WLANs have gained strong popularity in a number of vertical markets, including the health-care, retail, manufacturing, warehousing, and academic arenas. These industries have profited from the productivity gains of using hand-held terminals and notebook computers to transmit real-time information to centralized hosts for processing.

Key Words

academic	学术的，大学的，学会的
adapter	适配器
arena	舞台，场地，领域
baseband	基带
bounce	跳，反弹
brick	砖，砖块
carrier	载体，运输公司
ceiling	天花板，最高限度

cleverly	巧妙地，聪明地，灵敏地
demodulate	解调
divergent	分歧的，歧义的
dramatically	戏剧性地，引人注目地
electromagnetic	电磁的，电磁波
equalizer	均衡器
interference	干扰，干涉
massively	大范围地，强有力地，厚实地
metal	金属，合金
mobility	移动性，活动性
multiplexing	复用，倍增
orthogonal	正交的，互相垂直的
panel	面板，嵌板，控制板
popularity	普及，流行
prefix	前缀，字首
proponent	支持者，提倡者
propose	提议，建议
retail	零售商
risky	危险的，冒险的
router	路由器
unravel	拆开，解开
unsightly	不好看的，难看的
videoconference	视频会议，可视会议

Notes

[1] However, if an unsecured network has a stronger signal than a security-enabled one, it is safer for your data if you connect to the security-enabled network.

说明：本句有两个由"if"引导的条件状语从句。

译文：但是，当不安全网络中的信号比启用了安全保护的网络中的信号更强时，为了使数据更安全，最好还是连接到启用了安全保护的网络。

[2] To improve the signal strength, you can move your computer closer to the wireless router or access point, or move the router or access point so it is not close to sources of interference such as brick walls or walls that contain metal support beams.

说明："To improve the signal strength"是目的状语，有两个并列的主句，即"you can move your..."和省略主语的"move the router..."。

译文：若要提高信号强度，可以将计算机移动到距离无线路由器或访问点更近的位置，或者将路由器或访问点移动到远离干扰源（如砖墙或包含金属支撑梁的墙体）的位置。

[3] The information sent over such networks is encrypted, which can help protect your computer from unauthorized access.

说明:"which"引导的是非限定性定语从句。

译文:通过此类网络发送的信息是经过加密的信息,这有助于保护计算机免受未经授权的访问。

[4] As these signals proceed to a certain point, such as a wireless antenna, they often bounce and echo off objects, walls, furniture and floors, and arrive at the antenna at different times due to the different path lengths.

说明:"As these signals proceed to a certain point"是状语,"such as a wireless antenna"是"a certain point"的同位语,主句是并列句结构。

译文:在这些信号到达某一点(如无线天线)的过程中,它们常常因物体、墙壁、家具和地板等而产生反射,由于路径长度不同,信号到达天线的时间也不同。

4.2.2 Text B

The objective of the 3G was to develop a new protocol and new technologies to further enhance the mobile experience. In contrast, the new 4G framework to be established will try to accomplish new levels of user experience and multi-service capacity by also integrating all the mobile technologies that exist.

In spite of different approaches, each resulting from different visions of the future platform currently under investigation, the main objectives of 4G networks can be stated in the following properties: ubiquity, multi-service platform and low bit cost.

Ubiquity means that this new mobile networks must be available to the user, anytime, anywhere. To accomplish this objective services and technologies must be standardized in a worldwide scale. Furthermore the services to be implemented should be available not only to humans as have been the rule in previous systems, but also to everything that needs to communicate. In this new world we can find transmitters in our phone to enable voice and data communications, in our wrist, to monitor our vital signs, in the packages we send, so that we always know their location, in cars, to always have their location and receive alerts about an accident, in remote monitor/control devices, in animals to track their state or location, or even in plants. Based on this view, NTT DoCoMo (DO Communications Over Mobile Network), that has already a wide base of 3G mobile users, estimates the number of mobile communication terminals to grow in Japan from the actual 82.2 million to more than 500 million units by 2010.

A multi-service platform is an essential property of the new mobile generation, not only because it is the main reason for user transition, but also because it will give telecommunication operators access to new levels of traffic. Voice will loose its weight in the overall user bill with the raise of more and more data services. Low-bit cost is an essential requirement in a scenario where high volumes of data are being transmitted over the mobile network. With the actual price per bit, the market for the new high demanding applications, which transmit high volumes of data (e.g. , video), is not possible to be established. According to cost per bit should be between 1/10 and 1/100 of 3G systems.

To achieve the proposed goals, a very flexible network that aggregates various radio access technologies, must be created. This network must provide high bandwidth, from 50～100 Mbps for high mobility users, to 1Gbps for low mobility users, technologies that permit fast handoffs, an efficient delivery system over the different wireless technologies available, a method of choosing the wireless access from the available ones. Also necessary is a QoS framework that enables fair and efficient medium sharing among users with different QoS requirements, supporting the different priorities of the services to be deployed. The core of this network should be based in Internet Protocol version 6—IPv6, the probable convergence platform of future services. The network should also offer sufficient reliability by implementing a fault-tolerant architecture and failure recovering protocols.

Services also pose many questions as 4G users may have different operators to different services and, even if they have the same operator, they can access data using different network technologies. Actual billing using flat rates, time or cost per bit fares, may not be suitable to the new range of services. At the same time it is necessary that the bill is well understood by operator and client. A broker system would be advisable to facilitate the interaction between the user and the different service providers. Another challenge is to know, at each time, where the user is and how he can be contacted. This is very important to mobility management. A user must be able to be reached wherever he is, no matter the kind of terminal that is being used. This can be achieved in various ways, one of the most popular is the use of a mobile-agent infrastructure. In this framework, each user has a unique identifier served by personal mobile agents that make the link from users to Internet.

Key Words

alert	警觉的，注意的，提醒
bill	账单，清单，钞票
convergence	集合，收敛，会聚
deploy	调配，配置，配备
handoffs	切换，换手
infrastructure	基础结构，基本建设
reliability	可靠性
scenario	方案，剧情概要
sufficient	足够的，充分的，必然的
telecommunication	远程通信，无线通信，电信
tolerant	宽容的，容忍的
transmitter	发射器，传送者，传达者
ubiquity	无处不在，普遍存在，泛在性
vital	至关重要的，生死攸关的
wrist	手腕

4.2.3 Exercises

1. Translate the following phrases into English

(1)无线电频率

(2)电磁波

(3)实时信息

(4)控制面板

(5)延迟扩展

2. Translate the following phrases into Chinese

(1)Wireless LAN

(2)mobile user

(3)wireless network

(4)radio signal

(5)Coded Orthogonal Frequency Division Multiplexing

3. Identify the following to be True or False according to the text

(1)Wireless vendors now have a goal to boost wireless throughput beyond 600M bit/sec.

(2)In a wireless network, the computers are connected by wires instead of radio signals or cables.

(3)Public wireless networks are convenient, but if they are not properly secured, connecting to one might be risky.

(4)Using electromagnetic waves, WLANs transmit and receive data over the air, minimizing the need for wired connections.

(5)Ubiquity means that this new mobile networks must be available to the user, anytime, anywhere.

4. Reading Comprehension

(1)With current bit-rate technology, this ceiling tends to be around _____ bit/sec.

a. 10M to 20M

b. 10M to 25M

c. 15M to 20M

d. 15M to 25M

(2)In the list of available wireless networks, you will see a _____ that shows the wireless signal strength for each network.

a. signal

b. symbol

c. wire

d. word

(3)The new 4G framework to be established will try to accomplish new levels of user experience and _____ capacity by also integrating all the mobile technologies that exist.

a. multi-service

b. single-service

c. service

d. services

(4) While the 802.11a standard currently tops out at _____ in 20MHz channels, several firms are developing and proposing high-rate extensions to the 802.11a standard.

a. 48M bit/sec

b. 54M bit/sec

c. 56M bit/sec

d. 64M bit/sec

4.3 Distance Education

4.3.1 Text A

Distance education covers a multi-facetted techno-pedagogical reality, ranging from the simple decentralization of classroom activities to interactive multimedia models that make learning available whatever the time or the location. We now describe these various models and stress the principles that we have retained to design an integrated model of the Virtual Campus.[1]

The distance education world is bubbling. The rapidly evolving availability of multimedia telecommunication is giving way to a increasing number of techno-pedagogical models. We describe them in terms of six main paradigms:

♦ The enriched classroom

The enriched classroom is the place where technologies are used within a traditional setting in order to do a presentation, a demonstration or experimentation. It is a networked classroom allowing access to campus resources and external databases and it is sometimes called an "intelligent campus".

♦ The virtual classroom

The virtual classroom mainly uses videoconferencing to support distant learners and teachers, thereby recreating a telepresence type of classroom. Many university campus now have their own multimedia production studios so they may decentralize training at satellite locations.

♦ The teaching media

The teaching media is focused on the learner's workstation. It allows access to prefabricated multimedia course contents on CD ROM, either shipped by mail or available from a distant multimedia server. Instruction and didactic resources are offered in such a way that the learner can individualize his own learning process.

◆ Information highway training

Information highway training is also centered on the learner's workstation, which serves as a navigation and research instrument to find all kinds of useful educational information.[2] Essentially, a "Web course" is offered on a central site where instructions and pointers related to didactic resources (other Web sites) are gathered in order to accomplish learning activities.

◆ The communication network

The communication network uses the workstation, not only as a media support or as a way to access information, but also as a synchronous (desktop videoconferencing, screen sharing, etc) and as an asynchronous communication tool (electronic mail, computer teleconferencing, etc).[3]

◆ The performance support system

The performance support system concerns task-oriented training modules that are added to an integrated support system within a workplace.[4] Information has a "just-in-time" quality and training is seen as a process that is complementary and incorporated into the work process.

Each of these models has advantages and drawbacks. The first two are very popular at the present time. They rest on the traditional paradigm inherent in live information transmission: the teacher uses computerized and audiovisual equipment to animate a real-time multimedia group presentation, broadcast locally or to several distant locations where learners are gathered. This model requires costly equipment as well as the learners and teacher's physical presence simultaneously. Moreover, too often it reduces the learner's interaction and initiative to a level that is in no way better than that of a traditional course presentation in an auditorium.

This approach appears incapable of meeting the growing training needs in a socioeconomic context where lifelong learning, sought by busy and mobile people, involves cognitive abilities of a much higher level than what was required in the past.

The availability of Internet and multimedia technologies exposes the learner to numerous sources of information among which he must choices. The new paradigm (Figure 4-1) where the learner at the center of his learning process calls on many expertise sources, is better represented by the last four models described above than by the first two.

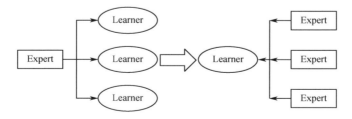

Figure 4-1　New paradigm

In the models of the "teaching media" or the "information highway training" and their application as it is currently, training is personalized but it is also deprived of an important collaborative dimension. However, this dimension can be reinstated if we use the computer as a tool to communicate.

Key Words

animate	活跃，有生气的，生气勃勃的
audiovisual	视听设备，视听的，音像的
auditorium	教室，讲堂
bubble	沸腾，起泡
cognitive	认识的，有认识力的，认知的
collaborative	合作的，协作的
complementary	互补的，补充的
costly	昂贵的，代价高的
decentralization	分散
deprive	剥夺，夺去
didactic	教导的，启发人的
expertise	专门技能，专家的意见
facetted	刻面，雕琢
initiative	主动性，自发的
lifelong	毕生的，终身的
navigation	导航，领航
pedagogical	教育学的，教授法的
prefabricate	预先构思
paradigm	范例，样式
reinstate	使恢复原来权力，使恢复原职，恢复
socioeconomic	社会经济学的
stress	强调，压力
techno	技术
telepresence	远程监控，临场

Notes

[1] We now describe these various models and stress the principles that we have retained to design an integrated model of the Virtual Campus.

说明：句中的"that we have retained"是定语，修饰"principles"；"to design an integrated model of the Virtual Campus"是目的状语。

译文：我们现在描述这些各种各样的模型并强调为设计虚拟校园的集成模型已保留的一些原则。

[2] Information highway training is also centered on the learner's workstation, which serves as a navigation and research instrument to find all kinds of useful educational information.

说明：句中的"which"引导非限定性定语从句，"to find all kinds of useful educational information"是目的状语。

译文：信息高速公路培训也集中在学习者工作站上，它用作导航和研究手段，以找到各种有用的教育信息。

[3] The communication network uses the workstation, not only as a media support or as a way to access information, but also as a synchronous (desktop videoconferencing, screen sharing, etc) and as an asynchronous communication tool (electronic mail, computer teleconferencing, etc).

说明：本句的"not only…but also…"结构是宾语的补足语。

译文：通信网络使用工作站，不仅作为媒体支持或作为访问信息的一种手段，而且也作为同步的（桌面视频会议、显示屏共享等）和异步的通信工具（电子邮件、计算机远程会议等）。

[4] The performance support system concerns task-oriented training modules that are added to an integrated support system within a workplace.

说明：句中的"that"引导定语从句，修饰宾语"modules"。

译文：性能支持系统涉及一些面向任务的培训模块，它们被添加到工作区内的集成支持系统。

4.3.2 Text B

Distance education has gained tremendous recognition for its ability to accept and use new educational technologies, which traditional education has been resistant to change and is not structured to make complete use of the new developments. The computer-based technologies now available for use in educational programs provide current and quality instructional options for teachers and students. Nontraditional students have been the chief beneficiaries of distance education and the increasing numbers of such students are prompting colleges and universities to evaluate the uses of technology to make education more accessible, efficient, and effective.

The definition of computer-mediated communication appears to parallel the definition of distance education in that it often removes the teacher from the student in time and location. Computer technologies have caused tremendous advancements in information storage and retrieval sciences that are combined with electronic communications to produce tremendous educational tools: Internet, telecommunication, electronic bulletin boards, electronic mail, video conferences, and many others.

Recent developments and decreasing costs are providing access to unsurpassed amounts of information from around the globe through electronic Internet. All that is required for students anywhere in the world to access this information is a computer, modem, telephone, and access port (commercial, educational, or business). The ability for students to conduct research without leaving their homes is changing the way that educational institutions structure research. Electronic publishing is one of the fastest growing fields, but establishment of standards before it will be accepted equally as printed material.

Telecommunications support distance education by providing delivery systems that carry programs and allow interaction between the participants. The availability of satellite and cable delivered programming provide economic options to school districts throughout the country. Programs such as the Jason Project and Space Explorer, allow any school to procure access to distance learning projects that can be seen world wide and interactive with each remote site. Rapid advancements in optical technology are providing methods of transferring increasing amounts of information and will allow remote students to use cable or telephone connections to use interactive multimedia programming.

Electronic bulletin boards and news services provide access to world wide discussions on any range of topics. These one-to-many communication platforms allow postings of assignments and course information for distance education. Group discussions allow individuals to analyze the thoughts of their peers as well as those of uncountable experts on any topic.

Increasing access to the Internet is promoting electronic mail (e-mail) as a one-to one or one-to many platform of communication. E-mail allows peer-to-peer conversations and can be almost instantaneous in response if the user sets up a two-way conversation mode. The ability to attach text and graphic files to e-mail allows the user to send papers and articles to any number of addresses. E-mail is extremely useful for correspondence between teacher and student, allowing feedback to any questions that the student might encounter while working on course material and allowing the teacher to transmit grades and feedback on submitted lessons.

Often, the need for face-to-face interaction is required and can easily be allowed through the use of video teleconferencing. Students enter into a cooperative learning process, enhancing the instructional experience and thus reducing isolation. Two-way full-motion video is available through satellite communications, coaxial cable, and in the near future through fiber-optic link. The current cost of video conferencing restricts its use to mostly commercial purposes, but as the price continues to drop it will undoubtedly be used throughout most distance education courses.

All of the technologies discussed above are easily controlled through computer-mediated communications and many are extensions of computer technology. There are many benefits derived from the use of computer-based communications. The first is the ability of computer programs to be interactive and provide feedback to the student. The second value is the ability of computers to become any and all existing media, including books and musical instruments. A third is that information can be presented from many different perspectives. Further values include the ability to use computers in simulation models and the ability to engineer computers to be reflective.

Key Words

beneficiary	受益人，受益者
bulletin	公告，公报
correspondence	一致，符合
district	地域，区域

establishment	企业，建立
extension	延长，延伸
institution	体系，机构
isolation	隔绝，隔离
nontraditional	非传统的
perspective	视角，观点
posting	任命，委派
procure	获得，取得
prompt	提示，鼓舞，推动
reflective	反射的，反映的
retrieval	检索
tremendous	巨大的，非常的
uncountable	无数的，不可数的
undoubtedly	无疑地，明确地

4.3.3 Exercises

1. Translate the following phrases into English

(1)网络教室

(2)远程教育

(3)信息高速公路

(4)教育信息

(5)虚拟校园

2. Translate the following phrases into Chinese

(1)electronic bulletin board

(2)performance support system

(3)intelligent campus

(4)virtual classroom

(5)task-oriented training module

3. Identify the following to be True or False according to the text

(1)Electronic bulletin boards and news services provide access to world wide discussions on any range of topics.

(2)Increasing access to the Internet is promoting e-mail as a one-to one platform of communication.

(3)Distance education covers a multi-facetted techno-pedagogical reality.

(4)Information highway training is not centered on the learner's workstation.

(5)Traditional students have been the chief beneficiaries of distance education.

4. Reading Comprehension

(1) The enriched classroom is the place where _____ are used within a traditional setting in order to do a presentation, a demonstration or experimentation.

    a. technologies
    b. news
    c. books
    d. discussions

(2) _____ has a "just-in-time" quality and training is seen as a process that is complementary and incorporated into the work process.

    a. Information
    b. Data
    c. Technology
    d. Service

(3) The availability of _____ and multimedia technologies exposes the learner to numerous sources of information among which he must choices.

    a. Extranet
    b. Internet
    c. Intranet
    d. Net

(4) E-mail is extremely useful for correspondence between _____ and student, allowing feedback to any questions that the student might encounter while working on course material and allowing the teacher to transmit grades and feedback on submitted lessons.

    a. student
    b. administrator
    c. manager
    d. teacher

4.4 Grid Computing

4.4.1 Text A

The grid infrastructure forms the core foundation for successful grid applications. This infrastructure is a complex combination of a number of capabilities and resources identified for the specific problem and environment being addresses.

In initial stages of delivering any grid computing application infrastructure, the developers/service providers must consider the following questions in order to identify the core infrastructure support required for that environment.[1]

- What problems are we trying to solve for the user?
- How do we address grid enablement simpler, while addressing the user's application

simpler?

• How does the developer help the user to be able to quickly gain access and utilize the application to best fit their problem resolution needs?

• How difficult is it to use the grid tools?

• Are grid developers providing a flexible environment for the intended user community?

• Is there anything not yet considered that would make it easier for grid service providers to create tools for the grid, suitable for the problem domain?

• What are the open standards, environments, and regulations grid service providers must address?

In general, a grid computing infrastructure component must address several potentially complicated areas in many stages of the implementation.

• Security

The heterogeneous nature of resources and their differing security policies are complicated and complex in the security schemes of a grid computing environment.[2] These computing resources are hosted in differing security domains and heterogeneous platforms. Simply speaking, our middleware solutions must address local security integration, secure identity mapping, secure access/ authentication, secure federation, and trust management. The other security requirements are often centered on the topics of data integrity, confidentiality, and information privacy.

The grid computing data exchange must be protected using secure communication channels, including SSL/ TLS and oftentimes in combination with secure message exchange mechanism such as WS-Security. The most notable security infrastructure used for securing grid is the grid security infrastructure (GSI). In most cases, GSI provides capabilities for single sign-on, heterogeneous platform integration and secure resource access /authentication. The latest and most notable security solution is the use of WS-Security standards. This mechanism provides message-level, end-to-end security needed for complex and interoperable secure solutions.

• Resource management

The tremendously large number and the heterogeneous potential of grid computing resources cause the resource management challenge to be a significant effort topic in grid computing environments. These resource management scenarios often include resource discovery, resource inventories, fault isolation, resource provisioning, resource monitoring, a variety of autonomic capabilities and service-level management activities. The most interesting aspect of the resource management area is the selection of the correct resource from the grid resource pool, based on the service-level requirements, and then to efficiently provision them to facilitate user needs.

It is important to understand multiple service providers can host grid computing resources across many domains, such as security, management, networking services, and application functionalities.[3] Operational and application resources may also be hosted on different hardware and software platforms, in addition to this complexity, grid computing middleware must provide efficient monitoring of resources to collect the required matrices on utilization, availability, and other information.

Another valuable and very critical feature across the grid computing infrastructure is found in the area of provisioning; that is, to provide autonomic capabilities for self-management, self-diagnosis, self-healing, and self-configuring. The most notable resource management middleware solution is the grid resource allocation manager (GRAM). This resource provides a robust job management service for users, which includes job allocation, status management, data distribution, and start/stop jobs.

- Information services

Information services are fundamentally concentrated on providing valuable information respective to the grid computing infrastructure resources. These services leverage and entirely depend on the providers of information such as resource availability, capacity, and utilization, just to name a few. These information services enable service providers to most efficiently allocate resources for the variety of very specific tasks related to the grid computing infrastructure solution. In addition, developers and providers can also construct grid solutions to reflect portals, and utilize meta-schedulers and meta-resource managers. These metrics are helpful in service level management in conjunction with the resource policies. This information is resource specific and is provided based on the schema pertaining to that resource. We may need higher level indexing services or data aggregators and transformers to convert these resource-specific data into valuable information sources for the end user.

The definition of grid computing resource sharing has changed based upon experience, with more focus now being applied to a sophisticated form of coordinated resource sharing distributed throughout the participants in a virtual organization. This application concept of coordinated resource sharing includes any resource available within a virtual organization, including computing power, data, hardware, software and applications, networking services, and any other forms of computing resource attainment.[4]

Key Words

aggregator	聚集，集合
authentication	鉴定，认证
confidentiality	秘密性，机密性
federation	同盟，联盟
heterogeneous	异种的，异类的
interoperable	互操作的
leverage	影响，杠杆作用
matrices（matrix 的复数）	矩阵
middleware	中间件
privacy	秘密，私下
provision	向……供应

regulation	规则，条例
scenario	方案，剧本
tremendously	惊人地，可怕地

Notes

[1] In initial stages of delivering any grid computing application infrastructure, the developers/service providers must consider the following questions in order to identify the core infrastructure support required for that environment.

说明：本句中的主语是"the developers/service providers"，"In initial stages of delivering any grid computing application infrastructure"是状语，"in order to"引导的是目的状语从句。

译文：在交付网格计算应用架构的起始阶段，为了确定特定环境要求的核心架构支持，开发商或服务提供商必须考虑以下一些问题。

[2] The heterogeneous nature of resources and their differing security policies are complicated and complex in the security schemes of a grid computing environment.

说明：本句主语有并列的两部分，即"The heterogeneous nature of resources and their differing security policies"，"complicated and complex"是表语，"a grid computing environment"是定语，修饰"security schemes"。

译文：在网格计算环境的安全方案中，资源的异构本质及其不同的安全策略是复杂的、复合的。

[3] It is important to understand multiple service providers can host grid computing resources across many domains, such as security, management, networking services, and application functionalities.

说明：本句中的"It"是形式主语，真正的主语是不定式结构"to understand…"，"such as security, management…"是"domains"的同位语。

译文：多种多样的服务提供商可以管理跨越许多领域的网格计算资源，如安全、管理、网络服务及应用功能等，了解这一点很重要。

[4] This application concept of coordinated resource sharing includes any resource available within a virtual organization, including computing power, data, hardware, software and applications, networking services, and any other forms of computing resource attainment.

说明：本句的"coordinated resource sharing"是定语，修饰主语；现在分词结构"including computing power…"是宾语补足语。

译文：协调资源共享的应用概念包括虚拟组织中的任何可获得的资源，包括计算能力、数据、硬件、软件及其应用、网络服务和其他任何计算资源获取的形式。

4.4.2 Text B

Geographic Information Systems（GIS）

A GIS, as defined by the National Science Foundation, is a computerized database management system used for the capture, storage, retrieval, analysis and display of spatial (e.g.

locationally defined) data.

A GIS consists of the following three parts:
- GIS software.
- Hardware.

Hardware needs to run a GIS depends on three interrelated variables: (1) Scope: the number of uses, number of applications and number of users; (2) Scale of data: the more detailed the maps, the more powerful the hardware needed; and (3) Functionality: the number of functions or operations to be performed on the data and the complexity of the functions.
- Database, both internal and external.

A key in developing a GIS system is geocoding. Geocoding is the process of linking attribute data to maps. Street address geocoding is the foundation technology of business geographic. It is said that about 80% of business data has some type of geographic component.

Geocoding is much trickier to do well than it may appear. Looking up address in a directory is easy in concept but can fail due to weaknesses in software, the geocoding reference directory or the data addresses themselves.

GIS systems allow a series of maps to be overlaid onto one another. By viewing the combination of computerized maps, a retailer could immediately see where his sales are high or low and where his competitors are strong or weak.

GIS systems allow data to be accessed in a variety of ways. Most full-function GIS systems combine three basic types of capabilities: (1) presentation mapping; (2) using maps as an organizing tool; and (3) spatial analysis.

The areas GIS systems are helping businesses including: (1) real estate, (2) direct mail marketing, (3) insurance, (4) banking, (5) service providers, (6) manufacturing, (7) transportation and distribution, and (8) retailing.

Analysis say the GIS market is being driven by several factors:
- Cheaper, faster hardware.

PC prices are dropping rapidly and desktop computers are becoming more powerful.
- Improved GIS software.

GIS software today is more friendly, allowing users who have no experience with GIS to quickly learn the systems.
- Reduced GIS prices.
- Greater variety of demographic data.
- Wide GIS use by businesses.

There are two kinds of GIS applications on the market. "Open Systems" allow direct import of data from your spreadsheet or database programs; "Closed Systems" do not. In general, open systems offer more utility. But they are more challenging to use. The onus is on the user to prepare the attribute data for import into the GIS, closed systems are easier to use. The attribute data arrives on your doorstep neatly bundled.

The GIS design process involves four basic elements: Geographic data; attribute data, both internal and external; mapping software, and hardware. Regarding geographical data, the first question to ask is, "What geographic area am I interested in?" Attribute data must be compatible with the GIS so it can be imported into the GIS. Mapping software should support data entry, data analysis, data output, data display, and data management. As for hardware, a GIS quickly will outgrow the minimum system requirements recommended by the GIS vendor. Go for more than you need at the present time.

The non spatial, internal and external attribute data is another major cost of developing a GIS. Buying external data such as commercial demographic databases and developing internal databases can account for as much as 80% of the total cost of a GIS system.

Key Words

attribute	属性，特征，价值
demographic	人口统计学的
full-function	全功能
geocoding	地理信息编码
import	进口，移入
neatly	整洁地，整齐地
onus	责任，义务
scale	规模
spatial	空间的，存在于空间的
utility	实用，功用，效用

4.4.3 Exercises

1. Translate the following phrases into English

(1)资源管理
(2)网格工具
(3)地理数据
(4)地理信息系统
(5)网格计算

2. Translate the following phrases into Chinese

(1)grid developer
(2)middleware solution
(3)secure message exchange mechanism
(4)real estate
(5)external attribute data

3. Identify the following to be True or False according to the text

(1)GIS systems don't allow a series of maps to be overlaid onto one another.

(2)GIS systems allow data to be accessed in a variety of ways.

(3)There are three kinds of GIS applications on the market.

(4)The grid infrastructure forms the core foundation for successful grid applications.

(5)It is important to understand multiple service providers can host grid computing resources across many domains.

4. Reading Comprehension

(1)The other _____ requirements are often centered on the topics of data integrity, confidentiality, and information privacy.

a. data

b. resource

c. privacy

d. security

(2)Another valuable and very critical feature across the _____ infrastructure is found in the area of provisioning.

a. grid computing

b. database

c. data structure

d. software and hardware

(3)_____ should support data entry, data analysis, data output, data display, and data management.

a. Hardware

b. Mapping software

c. Database

d. Geographic data

(4)The non spatial, _____ attribute data is another major cost of developing a GIS.

a. internal

b. internal and external

c. external

d. special

4.5 Electronic-Commerce

4.5.1 Text A

Electronic-commerce (EC) is doing business through electronic media. It means using simple, fast and low-cost electronic communications to transact, without face-to-face meeting between the two parties of the transaction.

From a service perspective, EC is a tool that addresses the desire of firms, consumers, and management to cut service costs while improving the quality of goods and increasing the speed of service delivery.[1]

High-speed network makes geographical distance insignificant. Businesses can sell goods to customers outside traditional markets, explore new markets and realize business opportunities more easily. Businesses can maintain their competitive advantage by establishing close contact with their customers and consumers at anytime through Internet by providing the latest information on products and services round the clock[2]. Internet provides companies with many markets in the cyberworld and numerous chances for product promotion. Besides, relationships with buyers can also be enhanced. By the use of multimedia capabilities, corporate image, product and service brand names can be established effectively through the Internet.

Classification of the EC field by the nature of the transactions

A common classification of EC is by the nature of transaction. The following types are distinguished:

Business-to-business (B2B). Most of EC today is of this type. It is the electronic market transactions between organizations.

Business-to-consumer (B2C). There are retailing transactions with individual shoppers. The typical shopper at Amazon.com is a consumer, or customer.

Consumer-to-consumer (C2C). In this category, consumer sells directly to consumers. Examples are individuals selling in classified ads and selling residential property, cars, and so on. Advertising personal services on the Internet and selling the knowledge and expertise is another example of C2C. Service auction sites allow individuals to put items up for auctions finally, many individuals are using Intranets and other organizational internal networks to advertise items for sale or services.

Consumer-to-business (C2B). This category includes individuals who sell products or services to organizations, as well as individuals who seek sellers, interact with them, and conclude a transaction.

Internet electronic-commerce security

There are numerous threats to the security of Internet electronic-commerce. Security breaches are most frequently discussed in terms of the Internet and the danger that hackers will intercept messages, misuse the information or modify the content of the message. The Internet is only one potential source of insecurity; further elements of the problem are:

The customer side: A customer can be impersonated, with or without the use of the customer's equipment. The use of stolen credit card details is the simplest example.

The vendor site: The vendor can trade inappropriately or dishonestly. Problems can range from customer details being stolen from the vendors files to bogus traders who set-up online and take money with no intention of supplying the advertised goods or services.

The security issues, across the network and at both ends, fall into a number of categories.

1. Confidentiality

When a message is sent electronically, the sender and receiver may desire that the message remain confidential, and thus not be read by any other parties. Thus the message must be made un-interpretable to everyone except the designated receivers, so as to give an electronic message the property of confidentiality.[3]

2. Authentication

When an electronic message is received by a user or a system, the identity of the sender needs to be verified in order to determine if the sender is who he claims to be.[4] To identify a user, at least one of the following types of information is generally required: something you have (e.g., a token); something you know (e.g., a PIN); or something you are (e.g., fingerprints or signatures).

3. Integrity

A message that has not been altered in any way, either intentionally or unintentionally, is said to have maintained its integrity. For electronic commerce, verifying that the order details sent by the purchaser have not been altered is one major security concern. Trading partners electronically sharing design specifications need assurance that the design specifications sent by the customer to their supplier, or vice versa, have not been altered in anyway during their electronic transmission.

4. Non-Repudiation

The term repudiation means to refuse to accept as having rightful authority or obligation, as in refusing to pay a debt because one refuses to acknowledge that the debt exists. For business transactions, unilateral repudiation of a transaction by either party is unacceptable and can result in legal action. Well-designed electronic commerce systems provide for non-repudiation, which is the provision for irrefutable proof of the origin, receipt, and contents of an electronic message. Companies engaged in electronic commerce are often vulnerable to non-repudiation risks.

Key Words

assurance	保证，担保，确信
auction	拍卖，标售
bogus	伪造的，假的
brand	商标，品牌
confidentiality	私密性，机密性
credit	信用，信誉

customer	客户,买主
dishonestly	不诚实地,不可靠地
fingerprint	指纹
hacker	黑客
impersonate	模仿,假冒
insignificant	微不足道的,不重要的
integrity	完整,完全
intentionally	有意地,故意地
intercept	拦截,截取,拦住
interpretable	可解释的,能说明的
irrefutable	无可辩驳的,驳不倒的
legal	法律的,合法的
misuse	误用,滥用
obligation	契约,债务,合约,责任
opportunity	机会,机遇
promotion	促进,推广,提升,晋升
receipt	收据,收条
repudiation	抵赖,拒绝,推翻
residential	住宅的,与居住有关的
rightful	正当的,合法的
shopper	顾客,购物者
signature	签名,署名,签署
unilateral	一方的,单方面的,单边的,片面的
unintentionally	无意地,非故意地
vulnerable	脆弱的,易受伤的

Notes

[1] From a service perspective, EC is a tool that addresses the desire of firms, consumers, and management to cut service costs while improving the quality of goods and increasing the speed of service delivery.

说明:"From a service perspective"是状语,主句是"EC is a tool","that"后面的部分是定语修饰"tool"。

译文:从服务的角度来看,电子商务是一种工具,可以满足企业、消费者和管理层在提高商品质量和加快服务交付速度的同时降低服务成本的愿望。

[2] Businesses can maintain their competitive advantage by establishing close contact with their customers and consumers at anytime through Internet by providing the latest information on products and services round the clock.

说明:"latest information"的意思是"最新信息",round the clock 的意思是"24 小时",本句的主句是"Businesses can maintain their competitive advantage","by"引导方式状语。

译文：通过在互联网上全天候地提供产品及服务的最新信息，商家可以与客户、消费者随时建立紧密联系来确保其竞争优势。

[3] Thus the message must be made un-interpretable to everyone except the designated receivers, so as to give an electronic message the property of confidentiality.

说明：本句的"so as to..."是目的状语从句。

译文：因此除指定的接收者外，传输的信息必须是无法解读的，这样才能赋予电子信息的私密性。

[4] When an electronic message is received by a user or a system, the identity of the sender needs to be verified in order to determine if the sender is who he claims to be.

说明："When"引导的是时间状语从句，"in order to"引导的是目的状语。

译文：在系统或用户收到一条电子信息时，发送者的身份必须能够得到验证，以确保发送者身份的可靠性。

4.5.2　Text B

Advertising is an attempt to disseminate information in order to effect a buyer-seller transaction. The Internet redefined the meaning of advertising. The Internet has enabled consumers to interact directly with advertisers and advertisements. In interactive marketing, a consumer can click with his or her mouse on an ad for more information or send an e-mail to ask a question. The Internet has provided the sponsors with two-way communication and e-mail capabilities, as well as allowing the sponsors to target specific groups on which they want to spend their advertising dollars, which is more accurate than traditional telemarketing. Finally, the Internet enables a truly one-to-one advertisement.

(1) Banners

Banner advertising is the most commonly used form of advertising on the Internet. As you surf your way through the information superhighway, banners are everywhere. The file size of the image should be about 7KB to 10KB. The smaller the file size, the quicker it loads. Designers of banners pay a lot of attention to the size of the image because long downloading times may cause a viewer to become impatient and move on before the banner is fully displayed. Typically, a banner contains a short text or graphical message to promote produce. Advertisers go to great lengths to design a banner that catches consumers' attention.

With the progress of Internet programming we are starting to find banners with video clips and sound. Banners contain links that, when clicked on, transfer the customer to the advertiser's home page. There are two types of banners: keyword banner and random banner. Keyword banners appear when a predetermined word is queried from the search engine. It is effective for companies who want to narrow their target audience. Random banners appear randomly. Companies that want to introduce their new products use random banners.

A major advantage of using banners is the ability to customize them to the target audience. One can decide which market segments to focus on. Banners can even be customized to one-to-one targeted advertisement. Also, "forced advertising" marketing strategy is utilized, which means customers are forced to see it. The disadvantages are high overall cost. If a company demands a successful marketing campaign, it will need to allocate a large percentage of the advertising budget to acquire a high volume of CPM.

There are several different forms of placing banner advertising on the Internet on others' Web sites. The most common forms are: Banner Swapping, Banner Exchanges, and Paid Advertising. Banner swapping means that company A agrees to display a banner of company B in exchange for company B displaying company A's ad. It is a direct link between Web sites. This is probably the least expensive form of banner advertising to establish and maintain, but it is also difficult to arrange. Frequently banner swapping does not work because a match is not possible. If there are several companies involved, however, a multi-company match may be easier to find. Banner exchange organizations arrange for a trading of three or more partners. Paid advertisement means purchasing banner ad space on the Internet. It is similar to buying ad space in other media.

(2) Splash screen

A splash screen is an initial Web site page used to capture the user's attention for a short time as a promotion or lead-in to the site home page or to tell the user what kind of browser and other software they need to view the site. The major advantage of a splash page over any other advertising method is that one can create innovative multimedia effects or provide sufficient information for a delivery in one visit.

(3) Spot leasing

Search engines often provide space (spot) in their home page for any individual business to lease. The duration of the lease depends upon the contract agreement between the Web site host and the lessee. Unlike banners, which show up at various times, the ad place on the spot will always be there; hence, competition is reduced. The disadvantage of spot leasing is that the size of the ad is often small and limited, causing some viewers to miss the ad. Also, the cost can be very high.

(5) E-mail

Another way to advertise on the Internet is to purchase e-mail addresses and send the company information to those on the list. The advantages of this approach are its low cost and the ability to reach a wide variety of targeted audiences. Most companies develop a customer database to whom they send e-mails. E-mail is emerging as a marketing channel that affords cost-effective implementation and better, quicker response rates than other advertising channels. A list of e-mail addresses can be a very powerful tool because you are targeting a group of people you know something about.

Key Words

advertiser	广告商
aggress	攻击，侵略

attention	注意力，注意
banner	旗帜广告，横幅广告
campaign	战役，运动
capture	夺取，捕获，占领
disseminate	传播，普及，散播
impatient	急躁的，急切的，渴望的
information superhighway	信息高速公路
lease	出租，租借
lessee	承租人，租户
predetermine	预定，先定，注定
segment	段，部分
sponsor	主办人，发起者，保证人
surf	冲浪，浏览，漫游
telemarketing	电话销售，电话营销

4.5.3 Exercises

1. Translate the following phrases into English

(1)电子商务
(2)传统市场
(3)电子媒介
(4)搜索引擎
(5)旗帜广告

2. Translate the following phrases into Chinese

(1)auction site
(2)trading partners
(3)electronic message
(4)buying ad space
(5)paid advertising

3. Identify the following to be True or False according to the text

(1)Electronic-commerce does business through electronic media.
(2)Internet provides companies with many markets in the cyberworld.
(3)With Electronic-commerce, you should not consider the network security.
(4)Companies engaged in electronic-commerce are often vulnerable to repudiation risks.
(5)Advertising is an attempt to disseminate information in order to effect a buyer-seller transaction.

4. Reading Comprehension

(1)_____includes individuals who sell products or services to organizations.

a. B2B

b. C2B

c. C2C

d. B2C

(2) When a message is sent electronically, the sender and receiver may desire that the message remain _____, and thus not be read by any other parties.

a. open

b. important

c. critical

d. confidential

(3) The major advantage of a _____ over any other advertising method is that one can create innovative multimedia effects or provide sufficient information for a delivery in one visit.

a. banner advertising

b. E-mail

c. splash page

d. spot leasing

(4) Security breaches are most frequently discussed in terms of the Internet and the danger that _____ will intercept messages, misuse the information or modify the content of the message.

a. hackers

b. sellers

c. customers

d. consumers

4.6 Neural Networks and Expert System

4.6.1 Text A

An Artificial Neural Network (ANN) is a mathematical or computational model for information processing based on a connectionist approach to computation. The original inspiration for the technique was from examination of bioelectrical networks in the brain formed by neurons and their synapses.

Structure of Neural Network

A neural network is a set of processing units (nodes) joined by links through which they communicate. Each unit is characterized by its activation state which changes in time. From the current unit's activation state a signal it sends into the network is calculated. This signal is carried over the links to other nodes. During the transmission it may be weakened or strengthened, depending upon the link's characteristics. Signals reaching a unit from its neighbours are combined into an input signal, from which the next activation state of that unit is computed.

Neural networks comprise following elements.
- Set of nodes
- Links
- Rules for calculating the input signal
- Activation function
- Output signal function

A typical feed-forward neural network is a set of nodes. Some of these are designated input nodes, some output nodes, and in-between are hidden nodes.[1] Each connection between neurons has a numerical weight. When the network is in operation, a value will be applied to each input node—the values being fed in by a human operator, from environmental sensors, or from some external program. Each node then passes its given value to the connections leading out from it, and on each connection the value is multiplied by the weight associated with that connection. Each node in the next layer then receives a value which is the sum of the values produced by the connections leading into it, and in each node a simple computation is performed on the value—a sigmoid function is typical. This process is then repeated, with the results being passed through subsequent layers of nodes until the output nodes are reached.[2]

Typically the weights in a neural network are initially set to small random values. This represents the network knowing nothing; its output is essentially a random function of its input. As the training process proceeds, the connection weights are gradually modified according to computational rules specific to the learning algorithm being used.[3] Ideally the weights eventually converge to values allowing them to perform a useful computation.

Types of Neural Network

- Single-layer perceptron

The earliest kind of neural network is a single-layer perceptron network, which consists of a single layer of output nodes; the inputs are fed directly to the outputs via a series of weights. In this way it can be considered the simplest kind of feed-forward network.

- Multi-layer perceptron

This class of networks consists of multiple layers of computational units, usually interconnected in a feed-forward way. Each neuron in one layer has directed connections to the neurons of the subsequent layers. In many applications the units of these networks apply a sigmoid function as an activation function.

Multi-layer networks use a variety of learning techniques, the most popular being back-propagation. Here the output values are compared with the correct answer to compute the value of some predefined error-function. By various techniques the error is then fed back through the network. Using this information, the algorithm adjusts the weights of each connection in order to reduce the value of the error function by some small amount. After repeating this process for a sufficiently large number of training cycles the network will usually converge to some state where the error of the calculations is small.[4] In this case one says that the network has learned a certain target function.

Advantages of Neural Network

ANN has several advantages, because it resembles the principles of the neural system structure.
- Learning: ANN has the ability to learn based on the so called learning stage.
- Auto organization: an ANN creates its own representation of the data given in the learning process.
- Tolerance to faults: because ANN stores redundant information, partial destruction of the neural network does not damage completely the network response.
- Flexibility: ANN can handle input data without important changes like noisy signals or others changes in the given input data.
- Real time: ANN is a parallel structure, if it is performed in this way using computers or special hardware real time can be achieved.

Objectives of Neural Network

Although various neural network models to a greater or a lesser extent remind one of the actual neural network, nevertheless they do differ from it (apart from cases where such models are constructed with the intention of an exact simulation). The purpose of the limits imposed is to simplify the reasoning, and to enable the execution of algorithms devised. ANNs enable the execution of intelligent operations, e.g. associations. The analysis of these networks also provides interesting conclusions on parallel processing.

Neural cells are characterized by a very long response time, but because of the massively parallel processing man is capable of instant execution of tasks for which conventional sequential machines require significantly more time; for instance, image recognition. Investigations of such processes will eventually lead to the design of fast algorithms that use the parallelism of artificial neurons. Parallel processing is especially important in building search algorithms employing neural networks.

Key Words

bioelectrical	生物电的
connectionist	连接
converge	收敛
inspiration	灵感

massively	大规模地,大量地
neuron	神经细胞,神经元
parallelism	平行,并行
perceptron	感知器
propagation	传播
redundant	冗余的,过多的
sigmoid	S形,S状弯曲
synapse	突触,(神经元)的突触
tolerance	容忍,忍受

Notes

[1] Some of these are designated input nodes, some output nodes, and in-between are hidden nodes.

说明:这是一个并列句,共有三部分,表语分别是"designated input nodes""output nodes"和"hidden nodes"。

译文:这些节点中的一些节点被指定为输入节点,另一些节点被指定为输出节点,在这两者之间的是隐藏节点。

[2] This process is then repeated, with the results being passed through subsequent layers of nodes until the output nodes are reached.

说明:本句的主句是"This process is then repeated",后跟的"with the results…"是状语。

译文:把计算结果传给后续层的节点,重复这个过程直至到达输出节点。

[3] As the training process proceeds, the connection weights are gradually modified according to computational rules specific to the learning algorithm being used.

说明:"As the training process proceeds"是状语,"according to…"是分词结构作为状语。

译文:随着训练过程不断进行,连接点的权值根据特定的学习算法的计算规则逐渐改变。

[4] After repeating this process for a sufficiently large number of training cycles the network will usually converge to some state where the error of the calculations is small.

说明:本句中的"After"引导的是时间状语,"where the error of the calculations is small"是定语,修饰"some state"。

译文:这些操作重复了相当多个训练周期后,网络通常收敛到某个状态,该状态下计算的误差很小。

4.6.2　Text B

Expert system

The reliance on the knowledge of a human domain expert for the system's problem solving strategies is a major feature of expert systems. An expert system is a set of programs that manipulate encoded knowledge to solve problems in a specialized domain that normally requires human expertise. An expert system's knowledge is obtained from expert sources and coded in a form suitable for the system to use in its inference or reasoning processes.

The expert knowledge must be obtained from specialists or other sources of expertise, such as texts, journal articles, and data bases. This type of knowledge usually requires much training and experience in some specialized field such as medicine, geology, system configuration, or engineering design. Once a sufficient body of expert knowledge has been acquired, it must be encoded in some form, loaded into a knowledge base, then tested, and refined continually throughout the life of the system.

Expert systems differ from conventional computer systems in several important ways.

1. Expert systems use knowledge rather than data to control the solution process. Much of the knowledge used is heuristic in nature rather than algorithmic.

2. The knowledge is encoded and maintained as an entity separate from the control program. As such, it is not compiled together with the control program itself. This permits the incremental addition and modification of the knowledge base without recompilation of the control programs.

3. Expert systems are capable of explaining how a particular conclusion was reached, and why requested information is needed during a consultation.

4. Expert systems use symbolic representations for knowledge and perform their inference through symbolic computations that closely resemble manipulations of natural language.

5. Expert systems often reason with meta-knowledge; that is, they reason with knowledge about themselves, and their own knowledge limits and capabilities.

The reasoning of an expert system should be open to inspection, providing information about the state of its problem solving and explanations of the choices and decisions that the program is making. Explanations are important for a human expert, such as a doctor or an engineer, if he or she is to accept the recommendations from a computer. Indeed, few human experts will accept advice from another human, let alone a machine, without understanding the justifications for it.

The exploratory nature of AI and expert system programming requires that programs be easily prototyped, tested, and changed. AI programming languages and environments are designed to support this iterative development methodology. In a pure production system, for example, the modification of a single rule has no global syntactic side effects. Rules may be added or removed without requiring further changes to the large program. Expert system designers often comment that easy modification of the knowledge base is a major factor in producing a successful program.

A further feature of expert systems is their use of heuristic problem-solving methods. As expert system designers have discovered, informal "tricks of the trade" and "rules of thumb" are an essential complement to the standard theory presented in textbooks and classes. Sometimes these rules augment theoretical knowledge in understandable ways; often they are simply shortcuts that have, empirically, been shown to work.

It is interesting to note that most expert systems have been written for relatively specialized, expert level domains. These domains are generally well studied and have clearly defined problem-solving strategies. Problems that depend on a more loosely defined notion of "common sense" are much more difficult to solve by these means. In spite of the promise of expert systems, it would be a mistake to overestimate the ability of this technology. Current deficiencies include:

- Difficulty in capturing "deep" knowledge of the problem domain.
- Lack of robustness and flexibility.
- Inability to provide deep explanations.
- Difficulties in verification.
- Little learning from experience.

In spite of these limitations, expert systems have proved their value in a number of important applications.

Key Words

empirically	以经验为根据地
exploratory	探索性
inability	无能，无能为力
incremental	增量的
inspection	检验，审查
justification	正当的理由
recommendation	推荐，介绍
recompilation	重新编译
refined	精炼的，精制的
resemble	像，相似
shortcut	捷径
specialist	专家
syntactic	句法学

4.6.3 Exercises

1. Translate the following phrases into English

(1)多层网络
(2)专家系统
(3)前馈神经网络
(4)输出信号函数
(5)误差函数

2. Translate the following phrases into Chinese

(1) artificial neural network

(2) activation state

(3) back-propagation

(4) training cycle

(5) human domain expert

3. Identify the following to be True or False according to the text

(1) Typically the weights in a neural network are initially set to small random values.

(2) Expert systems use data rather than knowledge to control the solution process.

(3) Neural cells are characterized by a very short response time.

(4) The earliest kind of neural network is a multi-layer perceptron network.

(5) The exploratory nature of AI and expert system programming requires that programs be easily prototyped, tested, and changed.

4. Reading Comprehension

(1) ANN can handle input data without important changes like _____ signals or others changes in the given input data.

a. input

b. output

c. noisy

d. parallel

(2) An expert system's knowledge is obtained from _____ and coded in a form suitable for the system to use in its inference or reasoning processes.

a. data sources

b. expert sources

c. databases

d. programs

(3) _____ use a variety of learning techniques, the most popular being back-propagation.

a. Expert systems

b. Single-layer networks

c. Multi-layer networks

d. Units

(4) Signals reaching a unit from its neighbours are combined into _____, from which the next activation state of that unit is computed.

a. an output signal

b. an input signal

c. signals

d. nodes

4.7 定语从句与状语从句

4.7.1 定语从句

定语从句又称关系从句，在句子中起定语作用，修饰一个名词或代词，有时也可修饰一个句子。被定语从句修饰的名词、词组或代词叫先行词，定语从句通常跟在先行词的后面。

例：This is the software that I would like to buy.

译文：这就是我想买的那个软件。

that I would like to buy 是定语从句，the software 是先行词。

通常，定语从句都由关系代词 that、which、who、whom、whose 和关系副词 when、where、why、how 引导。关系代词和关系副词往往放在先行词和定语从句之间，起联系作用，同时还代替先行词在句中担任一定的语法成分，如主语、宾语、定语和状语等。

例：The man who will give us a lecture is a famous professor.

译文：将要给我们讲演的人是位著名的教授。

在该句中，who will give us a lecture 是由关系代词 who 引导的定语从句，修饰先行词 the man，who 在从句中作为主语。

定语从句根据其与先行词的密切程度可分为限定性定语从句和非限定性定语从句。

1. 限定性定语从句

限定性定语从句与先行词的关系密切，是整个句子不可缺少的部分，没有它，句子的意思就不完整或不明确。这种定语从句与主句之间不用逗号隔开，译成汉语时，一般先译定语从句，再译先行词。

限定性定语从句修饰人，一般用关系代词 who，有时也用 that。若关系代词在句子中作为主语，则 who 用得较多，且不可省略；若关系代词在句子中作宾语，就应当使用宾格 whom 或 that，但在大多数情况下都可省略。若表示所属，就应用 whose。

限定性定语从句修饰物，用 that 较多，也可用 which。它们可在句中作为主语，也可作为宾语。若作为宾语，则大多可省略。

例：Microsoft Excel is a spreadsheet program that allows you to organize data, complete calculations, make decisions, graph data, develop professional looking reports, publish organized data on the Web, and access real-time data from Web sites.

译文：Microsoft Excel 是一个电子表格程序，用于组织数据、完成计算、做出决策、将数据图表化、生成专业水准的报告、在 Web 上发布组织好的数据，以及在 Web 站点上存取实时数据。

本句的主句是"Microsoft Excel is a spreadsheet program"，而"that"后的部分都是定语，修饰"a spreadsheet program"。

例：PCTOOLS are tools whose functions are very advanced.

译文：PCTOOLS 是功能很先进的工具。

因为 functions 和 tools 之间是所属关系，故用所有格 whose。

例：Those who agree with me please put up your hands.

译文：同意我的观点的人请举手。

who agree with me 是定语从句，修饰 Those。who 既是引导词，又在句中作为主语，who 不能省略。

例：Mouse is an instrument which operators often use.

译文：鼠标是操作员经常使用的一种工具。

which 引导的定语从句修饰 an instrument。因 which 在从句中作为 use 的宾语，故可省略。下面各例也是限定性定语从句。

例：That is the reason why I am not in favor of the plan.

译文：这就是我不赞成该计划的原因。

Potential energy is the energy that a body has by virtue of its position.

译文：位能是指物体由于自身的位置而具有的能量。

例：Generally speaking, much of the information needed by managers who occupy different levels and who have different responsibilities is obtained from a collection of existing information systems (or subsystems).

译文：一般而言，不同级别、不同职能的管理者所需的信息大多来自现有的信息系统（或子系统）集。

2. 非限定性定语从句

非限定性定语从句与先行词的关系比较松散，从句只对先行词附加说明，如果缺少，不会影响句子的主要意思。从句与主句之间常用逗号隔开，译成汉语时，从句常单独译成一句。非限定性定语从句在修饰人时用 who、whom 或 whose，修饰物时用 which，修饰地点和时间时用 where、when 引导。关系代词 that 和关系副词 why 不能引导非限定性定语从句。

例：Unlike a traditional LAN, which requires a wire to connect a user's computer to the network, a WLAN connect computers and other components to the network using an access point device.

译文：与需要一根线把用户的计算机与网络连接的传统局域网不同，无线局域网使用一个接入点装置把计算机和其他组成部分连接到网络。

本句中的"Unlike a traditional LAN"是状语，"which"引导的是非限定性定语从句，修饰"a traditional LAN"，"using an access point device"是方式状语。

例：We do experiments with a computer, which helps to do many things.

译文：我们利用计算机做实验，这有助于做许多事情。

Which 引导的非限定性定语从句是对先行词 a computer 的说明。

例：The meeting will be put off till next week, when we shall have made all the preparations.

译文：该会议将推迟到下周，那时我们将做好一切准备。

例：They'll fly to America, where they plan to stay for 10 days.

译文：他们将飞往美国，计划在那里逗留 10 天。

例：Mechanical energy is changed into electric energy, which in turn is changed into mechanical energy.

译文：机械能转变为电能，而电能又转变为机械能。

4.7.2 状语从句

英语中的状语从句通常由从属连词和起连词作用的词组来引导，用来修饰主句中的动词、形容词、副词等。

状语从句可位于主句前，也可位于主句后；前置时，从句后常用逗号与主句隔开；后置时，从句前通常不使用逗号。状语从句在句子中作为状语时，可表示时间、原因、目的、结果、条件、比较、方式、让步和地点等不同含义。

1．状语从句的分类

（1）时间状语从句

引导时间状语从句的连词或词组很多，但可根据所表示时间的长短及与主句谓语动词行为发生的先后这两点去理解和区别。

这些连词或词组有：when（当……时候），as（当……时候，随着，一边……一边），while（在……期间），before（在……之前），after（在……之后），since（自……以来），until（till）（直到……才），as soon as（一……就），no sooner...than（刚一……就……），once（一旦），every time（每次）等。

例：After some years of disuse, the term was revived in the mid-1970s to describe the interactive use of word and text processing systems, which would later be combined with powerful computer tools, thereby leading to a so-called "integrated electronic office of the future".

译文：经若干年搁置后，该词在 70 年代中期再次被用来描述字和文本处理系统的交互使用，这种系统后来又与强有力的计算机结合导致所谓的"未来综合电子办公室"的出现。

本句的"After some years of disuse"是时间状语，主句用被动语态，"which"引导的是非限定定语，"thereby leading…"是结果状语。

例：Fonts, keyboard layouts, and input direction change when you select a new language, and keyboard layouts can be customized.

译文：当你选择一种新语言时，字体、键盘布局和输入方向均可变化，并且可以定制键盘布局。

这是一个由"and"连接起来的并列句，本句的"when"引导的是时间状语。

例：When a program allocates expanded memory pages, the EMM returns a handle to the requesting program.

译文：当一个程序装入扩展存储器页中时，EMM 就将一个标志回复给这个请求程序。

例：No sooner had the push button been pressed than the motor began to run.

译文：按钮一被按下，电机就开始运转了。

例：When a resource should be accessed, the system looks up the user and group in the access control list and grants or denies access as appropriate.

译文：当一个资源被请求访问时，系统会在访问控制清单中查找用户和群并适当授权访问或拒绝访问。

例：Once we have a sufficiently precise theory of the mind, it becomes possible to express the theory as a computer program.

译文：一旦我们有了一个足够准确的智能理论，就有可能把该理论表示为计算机程序。

例：Check the circuit before you begin the experiment.

译文：检查好线路再开始做实验。

例：When you use a wireless router or access point to create a home network, you trade wired connectivity for connectivity delivered via a radio signal.

译文：当你使用一个无线路由器或接入点去建立一个家庭网络时，便由有线连接转换到由无线电信号传送的连接。

（2）原因状语从句

引导原因状语从句的连词和词组有：because（因为），as（由于），since（既然，由于），now that（既然），in that（因为）等。其中前三个较常用，它们表示原因的正式程度依次为 because＞since≥as。当原因是显而易见的或已为人们所知时，就用 as 或 since。由 because 引导的从句如果放在句首且前面有逗号，则可以用 for 来代替。但如果不是说明直接原因，而是用多种情况加以推断，则只能用 for。

例：Because your connection is being transmitted "in the clear", malicious users can use "sniffing" tools to obtain sensitive information such as passwords, bank account numbers, and credit card numbers.

译文：由于你的连接被畅通无阻地传送出去，所以恶意用户能够使用"嗅探"工具来获得敏感信息，如密码、银行账号及信用卡号。

本句由"Because"引导原因状语从句，"such as"后面的部分作为"sensitive information"的同位语。

例：The memory unit is an essential component in any digital computer since it is needed for storing the programs that are executed by the CPU.

译文：任何一台数字计算机都需要存储 CPU 所执行的程序，因此，存储器是计算机最重要的部件之一。

例：He is absent today, for he is ill.

译文：他今天缺席是因为他生病了。

（3）目的状语从句

目的状语从句由 in order that（为了，以便），so that（为了，以便），that（为了），lest（以免，以防），for fear that（以免，以防）等引导。

例：That requires follow through and communication with all employees so that they understand the emphasis and importance senior management is placing on security.

译文：这需要后续通过与所有员工的沟通，使他们认识到高层管理人员在安全上布置的重点和重要性。

本句的"so that"引导目的状语从句。

例：He handled the instrument with care for fear that it should be damaged.

译文：他小心地操纵这台仪器，生怕损坏它。

例：They require information relating to events as they occur so that appropriate action can be taken to control them.

译文：他们需要与事件发生时相关的信息以便采取适当举措以控制。

（4）结果状语从句

引导结果状语从句的连词有：so that（结果，以致），so…that（如此……以致），such…that（这样的……以致）等。注意 so 后接形容词或副词，而 such 后跟名词。so 还可以与表示数量的形容词 many、few、much 及 little 连用，形成固定搭配。

例：This problem is very difficult that it will take us a lot of time to work it out.

译文：这道题很难，我们要用很多时间才能解出。

（5）条件状语从句

条件状语从句用来表示前提和条件。通常由以下连词引导：

if（如果），unless（除非），provided / providing that（假如），as long as（只要），in case（如果），on condition that（条件是……），suppose / supposing（假如）等。

例：If you fail to secure your wireless network, anyone with a wireless-enabled computer within range of your wireless access point can hop a free ride on the Internet over your wireless connection.

译文：如果你不能确保你的无线网络的安全，那么任何拥有一台能激活无线网络的计算机的人只要在你的无线接口范围之内就能通过你的无线连接免费上互联网。

本句中的"If"引导的是条件状语从句；"with a wireless-enabled computer within range of your wireless access point"是定语，修饰主语"anyone"。

例：If we are going to say that a given program thinks like a human, we must have some way of determining how humans think.

译文：如果我们说一个给定的程序能像人一样思考，我们则必须有一些方法能确定人是怎样思考的。

本句由"if"引导条件状语；"determining how humans think"是定语，修饰"way"。

例：A physical body will not tend to expand unless it is heated.

译文：除非受热，否则物体不会有膨胀的倾向。

例：If conference rooms in offices can be equipped with the necessary audiovisual facilities, travel time and money can be saved by holding a teleconference instead of a face-to-face conference.

译文：如果会议室内能安装所需的声像设备，通过举行远距离会议则可代替面对面的会议，可节约旅行所花费的时间和经费。

（6）比较状语从句

比较状语从句经常是省略句，一般都是省略了重复部分；省略之后不影响句意，反而使结构简练。部分比较状语从句还有倒装现象。

比较状语从句由下列连词引导：as…as（像……一样），than（比），not so（as）…as（不像……一样），the more…the more（越……越），as…so（正如……那样）等。

例：Electron tubes are not so light in weight as semiconductor devices.

译文：电子管的重量不如半导体器件那么轻。

例：He finished the work earlier than we had expected.

译文：他完成这项工作比我们预计的更早。

（7）方式状语从句

方式状语从句通常由 as（如同，就像），as if（as though）（好像，仿佛）等连词引导。

as 引导的方式状语从句常常是一个省略句。as if 和 as though 两者的意义和用法相同，引出的状语从句常是一个虚拟语气的句子，表示没有把握的推测或一种夸张的比喻。（just）as…so…引导的方式状语从句通常位于主句后，意为"正如……，就像"。

例：The earth itself behaves as though it were an enormous magnet.

译文：地球本身的作用就像一个大磁铁一样。

例：They completely ignore these facts as if they never existed.

译文：他们完全不理会这些事实就好像它们不存在一样。

（8）让步状语从句

让步状语从句表示在相反的（不利的）条件下，主句行为依然发生了。

引导让步状语从句的有：(al) though（虽然），even if（though）（即使），as（尽管），whatever（不管），however（无论怎样），no matter（how，what，where，when）（不管怎样，什么，哪里，何时），whether…or（不论……还是）等。

例：It is important to detect such flows, even if they are very slight, before the part is installed.

译文：在安装部件之前，即使变形很轻微，也必须探测出来。

例：Much as computer languages differ, they have something in common.

译文：尽管计算机语言之间各不相同，但它们仍有某些共同点。

（9）地点状语从句

引导地点状语从句常用的词有：where（在……地方，哪里），wherever（在任何地方），everywhere（每一……地方）等。

例：She found her pen where she had left them.

译文：她的笔是她在原来放笔的地方找到的。

2．状语从句的翻译方法简介

状语从句的常用翻译方法有顺译法、倒译法、转译法和缩译法。

（1）顺译法

一般的句子可以按照原文提供的顺序直接翻译。当表示目的、原因等的状语从句在主句之前出现时，直接按照原句语序翻译。如果这些状语出现在主句之后，可以将它们提前或者保持原句顺序，翻译在主句之后，对主句的意思起到补充说明的作用。

例：Whenever you need any specific information, you can search it by Internet.

译文：每当你需要任何专门的信息时，你都可以通过互联网搜索得到。

例：The Internet is so powerful that you can get various information through it.

译文：互联网是如此强大，以致你可以通过它获取各种各样的信息。

（2）倒译法

当原文中的时间状语和地点状语在主句后面时，必须倒译；当原文中的原因状语从句、条件状语从句和让步状语从句在主句后面时，一般也可以倒译；另外，当特殊比较从句在主句后面时，必须倒译。

例：Many businesses became aware of network when they bought an expensive laser printer and wanted all the PCs to print to it.

译文：当企业购买了一台昂贵的激光打印机，并希望其全部的计算机都能使用该打印机时，他们就想到了网络。

例：In the Start menu folder, find the shortcut to the program you want to start each time you start Windows, and drag it to the Startup folder.

译文：在启动菜单文件夹时，找到每次启动 Windows 时想要启动的程序的快捷方式，然后把它拖到启动文件夹中。

（3）转译法

当对原文的逻辑含义进行分析后，会发现 when、where 不再单纯地表示时间、地点，或者翻译成"当……"或"在……"不合适时，可以考虑把这些词表示成"如果"的意思。另外，当状语从句比较短，而关联词可以省略时，可以把状语从句翻译成并列成分，这样也使得句子比较紧凑。

例：Where the Hz is too small a unit, we may use the MHz.

译文：当将赫兹作为单位太小时，我们可以使用兆赫兹。

例：Our whole physics universe, when reduced to the simplest terms, is made up of two things: energy and matter.

译文：我们的整个物理世界，如果用简单的话来说，是由能和物质这两样东西组成的。

（4）缩译法

有些关联词，如"so...that..."在很多情况下，可以省略翻译，这样使汉语的译文很简练，对于这样的句子可以采用"缩译法"进行翻译。

例：Computers work so fast that they can solve a very difficult problem in a few seconds.

译文：计算机工作如此迅速，在几秒钟内就可以解决一个很难的题目。

4.8 习题答案与参考译文

4.8.1 第 4 单元习题答案

4.1 Computer Network Basics

1. Translate the following phrases into English

(1)电子邮件　　　　　　　　　electronic mail
(2)简单邮件传输协议　　　　　Simple Mail Transfer Protocol

(3)文件传输协议　　　　　　　　　File Transfer Protocol
(4)广域网　　　　　　　　　　　　Wide Area Network
(5)环网　　　　　　　　　　　　　ring network

2. Translate the following phrases into Chinese

(1)Internet service provider　　　　互联网服务提供商
(2)star network　　　　　　　　　　星形网络
(3)backup computer system　　　　　备用计算机系统
(4)Local Area Network　　　　　　　局域网
(5)World Wide Web　　　　　　　　　万维网

3. Identify the following to be True or False according to the text

F　T　T　F　T

4. Reading Comprehension

(1)b　ISP
(2)d　Internet Explorer
(3)a　bus network
(4)a　destinations

4.2　Wireless Network

1. Translate the following phrases into English

(1)无线电频率　　　　　　　　　　　radio frequency
(2)电磁波　　　　　　　　　　　　　electromagnetic wave
(3)实时信息　　　　　　　　　　　　real-time information
(4)控制面板　　　　　　　　　　　　Control Panel
(5)延迟扩展　　　　　　　　　　　　delay spread

2. Translate the following phrases into Chinese

(1)Wireless LAN　　　　　　　　　　　　　　　　无线局域网
(2)mobile user　　　　　　　　　　　　　　　　　移动用户
(3)wireless network　　　　　　　　　　　　　　无线网络
(4)radio signal　　　　　　　　　　　　　　　　无线电信号
(5)Coded Orthogonal Frequency Division Multiplexing　编码正交频分复用

3. Identify the following to be True or False according to the text

F　F　T　T　T

4. Reading Comprehension

(1)a　10M to 20M
(2)b　symbol
(3)a　multi-service

(4)b 54M bit/sec

4.3　Distance Education

1. Translate the following phrases into English

(1)网络教室　　　　　　　　　　　　networked classroom
(2)远程教育　　　　　　　　　　　　distance education
(3)信息高速公路　　　　　　　　　　information highway
(4)教育信息　　　　　　　　　　　　educational information
(5)虚拟校园　　　　　　　　　　　　virtual campus

2. Translate the following phrases into Chinese

(1)electronic bulletin board　　　　　　电子公告板
(2)performance support system　　　　性能支持系统
(3)intelligent campus　　　　　　　　智能校园
(4)virtual classroom　　　　　　　　　虚拟教室
(5)task-oriented training module　　　　面向任务的训练模型

3. Identify the following to be True or False according to the text

T　F　T　F　F

4. Reading Comprehension

(1)a　technologies
(2)a　Information
(3)b　Internet
(4)d　teacher

4.4　Grid Computing

1. Translate the following phrases into English

(1)资源管理　　　　　　　　　　　　resource management
(2)网格工具　　　　　　　　　　　　grid tool
(3)地理数据　　　　　　　　　　　　geographic data
(4)地理信息系统　　　　　　　　　　geographic information systems
(5)网格计算　　　　　　　　　　　　grid computing

2. Translate the following phrases into Chinese

(1)grid developer　　　　　　　　　　网格开发者
(2)middleware solution　　　　　　　　中间件解决方案
(3)secure message exchange mechanism　安全信息交换机制
(4)real estate　　　　　　　　　　　　房地产
(5)external attribute data　　　　　　　外部属性数据

3. Identify the following to be True or False according to the text

F T F T T

4. Reading Comprehension

(1) d security

(2) a grid computing

(3) b Mapping software

(4) b internal and external

4.5 Electronic-Commerce

1. Translate the following phrases into English

(1) 电子商务　　　　　　　　　electronic-commerce

(2) 传统市场　　　　　　　　　traditional market

(3) 电子媒介　　　　　　　　　electronic media

(4) 搜索引擎　　　　　　　　　search engine

(5) 旗帜广告　　　　　　　　　banner advertising

2. Translate the following phrases into Chinese

(1) auction site　　　　　　　　拍卖网站

(2) trading partners　　　　　　贸易伙伴

(3) electronic message　　　　　电子消息

(4) buying ad space　　　　　　购买广告空间

(5) paid advertising　　　　　　付费广告

3. Identify the following to be True or False according to the text

T T F F T

4. Reading Comprehension

(1) b C2B

(2) d confidential

(3) c splash page

(4) a hackers

4.6 Neural Networks and Expert System

1. Translate the following phrases into English

(1) 多层网络　　　　　　　　　multi-layer network

(2) 专家系统　　　　　　　　　expert system

(3) 前馈神经网络　　　　　　　feed-forward neural network

(4) 输出信号函数　　　　　　　output signal function

(5) 误差函数　　　　　　　　　error-function

2. Translate the following phrases into Chinese

(1) artificial neural network　　　人工神经网络
(2) activation state　　　　　　　激活状态
(3) back-propagation　　　　　　反向传播
(4) training cycle　　　　　　　　训练周期
(5) human domain expert　　　　人类领域专家

3. Identify the following to be True or False according to the text

T　F　F　F　T

4. Reading Comprehension

(1) c　noisy
(2) b　expert sources
(3) c　Multi-layer networks
(4) b　an input signal

4.8.2　第4单元参考译文

4.1　计算机网络基础

4.1.1　课文 A

网络建立了计算机之间的通信。当人们在不同的地方工作时，该系统是特别有用的。它提高了通信的速度和准确性，可防止信息被放错地方，且可自动确保关键信息的分发。

一个网络中包含若干台由通信线路连接起来的计算机。这些计算机能单独工作，但它们的活动也是相互协调的。将计算机连接起来的最初目的是在不同大型机用户间交换信息（程序、数据文件）。当更小的计算机出现，并且在一个小机构内部也有很多台机器时，把这些小的计算机连接起来以共享资源（打印机、磁盘、处理时间）也变得很吸引人了。把这些不同型号、不同大小的计算机连接起来，但又保留其独立性而产生的系统称为计算机网络。

局域网（LAN）是专有的通信网络，它可以覆盖一个有限的地域，如一个办公室、一幢建筑或一群建筑等。硬件资源共享可使网上的每台计算机访问并使用由于太昂贵而无法为每人配备的设备。信息资源共享允许局域网上每个计算机用户访问存储在网上其他计算机中的数据。广域网（WAN）相对于局域网，在覆盖的地理范围上要更大一些，它使用电话线、微波、卫星或这些通信信道的组合来传递信息。

通信网中设备的配置（或称物理布局）称为拓扑。通信网络通常被配置为三种模式中的一种，或它们的组合。这些配置是星形、总线和环形网络。虽然这些配置也可用于广域网，但在此仅在局域网中对它们进行说明。连接到网络上的设备，如终端、打印机或其他计算机，称为节点。

◆ 星形网络

星形网络由一台中央计算机和一台或多台连接到该中央计算机上并形成星形结构的终端或计算机组成。纯粹的星形网络仅由终端和中央计算机之间的点对点的连线组成，但是大多数星形网络由点对点的连线和多点连线组成。星形网络配置通常用于中央计算机中含有处理来自终端的输入请求所需要的全部数据的场合，如航空订票系统。如果查询是在星形网络上处理的，那么回答该查询所需的所有数据应该包含在存储在中央计算机的数据库中。

星形网络的效率相对较高，严密的控制可保证网上处理的数据的安全。其主要缺点是整个网络都依赖于中央计算机及其相关的硬件和软件。如果其中任何部分不能正常工作，整个网络就会瘫痪。所以，在大多数大型星形网络中都有一个备用的计算机系统，以防止主系统出现故障。

◆ 总线网络

使用总线网络时，网络中的所有设备都连接到同一根电缆上。信息可以从任何一台个人计算机向任何方向传给另一台计算机，任何信息都可以被传送到某一指定设备。总线网络的优点是设备可以在任何一点与网络连接或分离，而不会影响网络其他部分的工作。此外，如果网络上的某台计算机出现故障，不会影响网络上其他用户。

◆ 环形网络

环形网络不使用中央计算机，而是连接成环形来实现计算机之间的相互通信。当处理不是在中心位置而是在当地进行时，环形网络是非常有用的。例如，计算机可以放在三个部门：财务部门、人事部门和收发部门。这三个部门的计算机可以分别完成各部门所要求的处理。但是收发部门的计算机偶尔需要与财务部门的计算机通信，以更新存储在财务部门计算机上的某些数据。数据只能沿着环形网络的一个方向顺序通过每个节点进行传送。因此，环形网络的缺点是如果一个节点出现故障，由于数据不能通过出现故障的节点，就会使整个网络无法工作。环形网络的优点是所需的电缆线少，因此，网络的电缆费用较低。

4.1.2 课文 B

Internet 是一个由相互连接的政府、教育、商务计算机网络组成的国际互联网——实际上是一个由多网络组成的网络。人们使用相应的软件，通过计算机终端或者个人计算机在 Internet 上进行交流。他们将数据放进一个网际协议（IP）数据包中——一个电子信封，并附上该数据包要送到 Internet 上特定目的地的"地址"。介于起点和终点网络之间的通信软件"读到了"流经 Internet 数据包上的地址并将这些数据包转送至目的地。

如何使用互联网

Windows 为你提供了快速方便的互联网访问，无论你正在运行什么程序或执行什么任务。活动桌面允许你自定义工作区，地址栏帮助你从任何窗口连接到互联网。你还可以找到各种工具，帮助你与他人及其他计算机通信。

大多数人通过网络连接或互联网服务提供商（ISP）连接到互联网。ISP 向你提供了用于拨号登录到互联网服务器的服务号码。一旦你接入系统，就能访问互联网，使用 ISP 所提供的电子邮件及所有其他服务。ISP 也会向你提供详尽的信息，用于对你的计算机连接互联网进行配置。如果你使用了网络连接，系统管理员会向你提供这些信息。

使用包含在 Windows 中的通信工具，就能利用计算机发送电子邮件、处理电话呼叫、发送传真或举行视频会议。例如，你能使用电话拨号程序接听电话或参加公司的视频会议。通过 Internet Explorer 和互联网连接，可以在万维网上查找并浏览信息。可以在地址栏内输入想要访问的网页地址，或从收藏夹中单击某一地址。Internet Explorer 能让你在互联网上找人、查询某个企业及感兴趣的相关主题信息。

互联网服务

最流行和应用最广的互联网应用服务包括如下方面。

（1）万维网（WWW）

万维网是一种提供互联网服务的大型网络，它向运行客户应用程序，如浏览器软件的终端提供超文本和其他服务。万维网允许用户从动态链接信息的全球网络服务器系列中搜索、访问和下载信息。

（2）电子邮件（E-mail）

电子邮件让用户能撰写信函并将其发送给一个人或一群人。邮件应用程序的另一功能是允许用户阅读已经收到的信函。在互联网上有两种电子邮件协议：一个是简单邮件传输协议（SMTP），它接收传来的邮件并从中复制报文到相应的邮箱中；另一个邮件协议是邮政协议-3（POP3），它从远程邮箱中取出电子邮件并将其存储在用户本地机器上，以便于以后阅读。

（3）文件传输协议（FTP）

FTP 是 TCP/IP 协议栈中的应用协议，用来在网络节点间传送文件。TCP/IP 协议包含一种文件传输应用程序，它允许用户发送和接收任意规模的程序或数据文件。

（4）远程通信网（Telnet）

Telnet 用于远程终端连接，允许用户登录到远程系统上，并像连接到本地系统那样使用资源。

4.2 无 线 网 络

4.2.1 课文 A

在无线网络中，计算机通过无线电信号而不是电线或电缆进行连接。无线网络的优点是移动灵活和没有难看的电线。缺点是连接速度比有线网络的慢，并且会受到其他无线设备的干扰。计算机需要内部或外部无线网络适配器。若要查看计算机是否具有无线网络适配器，请执行下列操作：

通过单击"开始"按钮，依次单击"控制面板""网络和 Internet""网络和共享中心"，然后单击"管理网络连接"，单击"管理网络连接"，列出计算机中安装的适配器。

在可用的无线网络列表中，可以看到一个显示每个网络无线信号强度的符号。强信号通常意味着无线网络很近或者没有干扰。为了获得最佳性能，请连接到信号最强的无线网络。但是，当不安全网络中的信号比启用了安全保护的网络中的信号更强时，为了使数据更安全，最好还是连接到启用了安全保护的网络。若要提高信号强度，可以将计算机移动到距离无线路由器或访问点更近的位置，或者将路由器或访问点移动到远离干扰源（如砖墙或包含金属支撑梁的墙体）的位置。

公共无线网络非常方便，但是如果没有适当的安全保护，连接到公共无线网络可能很危险。尽可能只连接到要求网络安全密钥或具有某些其他安全形式（如证书）的无线网络。通过此类网络发送的信息是经过加密的信息，这有助于保护计算机免受未经授权的访问。

在可用的无线网络列表中，每个网络会标记为安全的或不安全的。如果确实要连接到不安全的网络，请注意拥有适当工具的用户可能会看到你执行的所有操作，其中包括访问的网站、处理的文档，以及使用的用户名和密码。连接到此类网络时，不应处理文档或访问包含个人信息（如银行记录）的网站。

无线设备供应商面临着应用的挑战，这些应用要求支持的带宽越来越大，如 IP 电话、流式视频和视频会议等。为了显著提高吞吐量，802.11a 的支持者必须解决室内射频这一重大挑战。他们必须开发出一种方法，解决目前 2.4GHz 单载波延迟扩展系统的延迟扩展问题。

延迟扩展是由被发射的射频的反射引起的。在这些信号到达某一点（如无线天线）的过程中，它们常常因物体、墙壁、家具和地板等而产生反射，由于路径长度不同，信号到达天线的时间也不同。因此，需要基带处理器或均衡器，"拆分"有歧义的射频信号。延迟扩展必须小于符号速率，或小于为传输而进行的数据编码的速率。如果不是这样，有些延迟的信号会扩散到下一个符号的传输中。这就给能被持续获得的最大位速率加了个上限。

在采用目前位速率技术的情况下，这个上限往往会在 10~20Mbps 之间。802.11a 标准通过一种称为编码正交频分复用（COFDM）的创新调制技术，巧妙地解决了这个挑战，COFDM 早已用于欧洲的数字电视和音频的传输。COFDM 通过下列两种方法突破了数据位速率的上限：(1) 以大规模并行的方式发送数据；(2) 放慢符号速率，致使每个符号的传输比典型的延迟扩展长得多。为了让所有的延迟信号在基带处理器对数据解调之前"确定下来"，在符号传输的起始处插入保护间隔（有时也称周期前缀）。

无线设备供应商现在的目标是将无线吞吐量提高到 100Mbps 以上。802.11a 标准目前能在带宽为 20MHz 的信道上达到最高 54Mbps，但已有几家公司正在开发和建议 802.11a 标准的高速率扩展。这些建议一般都是设想把吞吐量至少提高 1 倍，达到 108~155Mbps 之间的某个值。

对于局域网有一个新开发的网络：无线局域网。无线局域网使用电磁波通过空气传送和接收数据，减少了有线连接的需求。这样，无线局域网把数据连接和用户移动性结合起来，通过简化的配置，形成了移动的局域网。随着近几年的发展，无线局域网在一些市场领域已经获得了广泛的普及，其中包括保健、零售、制造业、仓储业和学术界。这些工业利用手提终端和笔记本电脑将实时信息传送到中央主机进行处理，从而从生产力增加中获益匪浅。

4.2.2 课文 B

3G 的目标是开发新的协议和新的技术，进一步增强移动感受。相比之下，4G 构架的建立将致力于完成高水平的用户体验，以及将现存的所有移动技术整合从而实现多业务能力。

对正处于研究中的未来通信平台，尽管不同观点引发了不同的发展途径，4G 网络的主要目标是立足于：无处不在、多业务平台、低比特成本。

无处不在意味着用户能在任何时间、任何地点获得新的移动网络服务。要达到这一目标，移动业务和技术的标准化必须具有全球性。而且，不仅是先前系统服务的人群，其他需要通信的所有人群都必须获得将要实施的业务。在这个新的世界上，我们会发现发射器存在于完成声音和数据通信的手机里，存在于手腕上监视我们健康状况的仪器内，存在于我们邮寄的包裹中，这样我们就可以知道包裹的所在位置。在汽车上安装发射器，就可以知道车的位置和收到其事故报警；在远距离的监视器或控制器上安装发射器，在动物身体上安装发射器可以跟踪它们的行迹；甚至在植物上也安装发射器。基于这些观点，有 3G 用户基础的 NTT DoCoMo（通过移动网络实现通信）公司估计，日本的移动通信终端数量将从现在的 8220 万台增加到 2010 年的 5 亿台。

新的移动技术必须具有多业务平台，这不仅仅因为业务平台是用户过渡的主要理由，也因为多业务平台让通信操作员接入新级别通信。在总的用户账单中，语音的权重将下降，数据通信越来越多。新方案中低比特成本是主要要求，因为在移动网中主要传输大量数据。新的高端应用市场不能接受以现行每比特的价格传输大量数据的高要求（如视频）。按成本，每比特的价格应该在 3G 系统的 1/10～1/100 之间。

要达到提议的目标，必须建立融合多种无线接入技术的灵活的网络，这个网络必须能为高速移动的用户提供 50～100Mbps 的带宽，为低速移动的用户提供 1Gbps 的带宽，技术上要能高速切换，允许在具有不同无线技术的系统间高效传输数据，从允许的方式中选择无线接入方式。另外，要求 QoS 体系能使不同 QoS 级别的用户公平高效地共享媒体，支持要实施的不同优先权的业务。这个网络的核心应该以 IPv6 协议为基础，它是未来业务可能融合在一起的平台。通过使用容错结构和纠错协议，该网络应该有足够好的可靠性。

4G 用户可能有不同的运营商提供不同的业务，这也提出许多业务方面的问题。即使是相同的运营商，用户也能利用不同的网络技术完成数据收发。在时间或每比特花费上，现在的账单使用统一费率，但这可能不适用于新一代的业务范围。同时，账单能被运营商和用户易于理解也是必须的。通过代理商减轻用户和不同业务运营商的相互关系也是必须的。另一个挑战是，随时要知道用户在哪里，如何与之联系。这对移动性管理是非常重要的。无论用户身在何处，他使用哪种类型的终端，网络必须能联络到他。这可以用多种途径实现，其中最流行的就是引入移动代理机制，在这个框架下，每个用户有个人移动代理提供的唯一标识，代理商提供用户和互联网间的连接。

4.3 远程教育

4.3.1 课文 A

远程教育覆盖了多方面的技术—教学法的现实，范围从简单的教室活动分散化到交互式多媒体模型，使得无论是在什么时间还是什么场所学习都可行。我们现在描述这些各种各样的模型并强调为设计虚拟校园的集成模型已保留的一些原则。

远程教育领域正在蓬勃发展。迅速发展的多媒体远程通信正使得技术—教学法模型的数量日益增加。我们借助 6 个主要范式描述它们。

◆ 内容丰富了的教室

内容丰富了的教室指在传统的环境中使用一些技术以进行展示、演示或实验。它是一个允许访问校园资源和外部数据库的网络化教室，有时称为"智能校园"。

◆ 虚拟教室

虚拟教室主要使用视频会议来支持远程的学习者和教师，因此，重新建立了远程出席类型的教室。很多大学校园现在有它们自己的多媒体产品演播室，因此它们可以在一些附属卫星教学点进行分散培训。

◆ 教学媒体

教学媒体集中在学习者工作站上。它允许访问预制在 CD ROM 上的多媒体课程内容，该内容既可以通过邮件传递，也可以从远程多媒体服务器获得。教育和教学用的资源是以这样一种方式提供的，使得学习者能个性化自己的学习过程。

◆ 信息高速公路培训

信息高速公路培训也集中在学习者工作站上，它用作导航和研究手段，以找到各种有用的教育信息。本质上，一门"Web 课程"在中心网站上提供，为完成一些学习活动在那里收集与教学用资源（其他 Web 网站）相关的指示和指针。

◆ 通信网络

通信网络使用工作站，不仅作为媒体支持或作为访问信息的一种手段，而且也作为同步的（桌面视频会议、显示屏共享等）和异步的通信工具（电子邮件、计算机远程会议等）。

◆ 性能支持系统

性能支持系统涉及一些面向任务的培训模块，它们被添加到工作区内的集成支持系统。信息具有"及时"品质并把培训看成一个补充的并被加入工作流程的过程。

这些模型中的每一个都有一些优缺点。前两者在目前十分流行。它们停留在现场信息传输中固有的传统范式：教师使用计算机化设备和视听设备来制作实时多媒体组演示文稿的动画，在本地广播或向学习者聚集的几个远距离场所广播。这种模型需要昂贵的设备，以及学习者和教师同时在场。而且，它经常把学习者的交互作用和主动积极性降低到与课堂中的传统讲课差不多的程度。

这种方法看来是不能满足社会经济环境中不断增长的培训需求的。在这种环境中，忙碌且流动的人们所寻求的毕生学习要包括比过去所需求的高得多的认识能力。

互联网和多媒体技术的可用性使学习者面临为数众多的信息源，他必须在其中做出选择。该新范式（如图 4-1 所示）（处于其学习过程之中心的学习者访问很多专业知识源）用上面描述的后四种模型表示比用前两种表示要好。

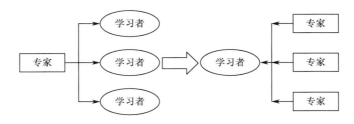

图 4-1 新范式

在"教学媒体"或"信息高速公路培训"模型和它们的应用中，按当前实际情况，培训被个性化，但是它也丧失了重要的协作方面。然而，如果我们使用计算机作为通信工具，协作方面能被恢复。

4.3.2　课文 B

由于远程教育能够接受和使用新的教育技术而得到了广泛的认可，然而传统教育还停滞不前，不能充分利用计算机的新发展。教育中可用的以计算机为基础的技术向教师和学生们提供了最新的和高质量的教育选择。屏弃传统教育的学生是远程教育的首要受益人，而且该学生队伍不断壮大促使高校进一步使用新技术，使教育更便捷、有效及有影响力。

计算机媒体交流的定义似乎与远程教育的定义类似，因为在远程教育中，教师和学生在时间与地点上往往不是共存的。计算机技术与电子通信技术结合在信息存储和检索科学方面取得了巨大的进展，从而产生了重要的教育工具：互联网、远程通信、电子公告板、电子邮件、视频会议等。

最新的发展和不断降低的费用提供了通过互联网从全球获得大量信息的途径。在世界任何地方的学生要获得信息所需要的只是一台计算机、一个调制解调器、电话和访问端口（商业的、教育的或商务的）。学生足不出户就可以进行研究的能力正在改变教育院校研究的方法。电子出版领域是发展最快的领域之一，但是在它作为印刷物而被广泛地接受之前需要制定一些标准。

远程通信通过提供传递系统支持远程教育，发送教育节目并使参与者之间互相沟通。卫星和光缆用来发送教育节目为全国各地的学校提供了经济便捷的途径。像杰森项目和太空探索者一类的项目就可以使任何学校接受世界范围内可以收到的远程教育项目，并且相隔很远的接收点之间可以沟通。光学技术的快速发展为传输日益增加的大量信息提供了方法，并且使相隔很远的学生通过电缆或电话连接来使用交互式多媒体程序。

电子公告板及新闻服务为在世界范围内讨论任意范围的题目提供了途径。这些一对多的通信平台使远程教育能够发送作业和课程信息。成组讨论使每个人能够就任何议题分析同学及不计其数的专家的见解。

互联网的日益普及正在促进电子邮件作为一对一或一对多的通信平台。E-mail 允许对等会话模式，而且如果用户建立了双向通话模式几乎同时就可以收到返回的信息。把文本或图片文件附加到 E-mail 上的能力使用户能把作业和论文发送到任意多的地址。E-mail 对于教师和学生间的通信极有用，它使学生在学习课程时，遇到的问题能及时得到反馈信息，还可以使教师传送当前课程的成绩并回答相关问题。

一般来说，视频电话会议系统能够方便地实现面对面的交流。学生进入互相合作的学习过程，增强了教学体验而减少了单独学习的弊端。双向电视图像可通过卫星通信、同轴电缆和在不久的将来要用的光纤链路来发送。目前，视频会议的造价限制了它在大多数商业领域的使用，但是随着价格的持续下降，它将毫无疑问地用于远程教育课程。

以上讨论的所有技术可以容易地由计算机中央通信来控制，而且许多是计算机技术的扩展。使用以计算机为基础的通信有许多益处。第一，通过计算机可以互相沟通，并提供给学生反馈信息。第二，计算机能够代替任何已存在的媒体，包括课本和乐器。第三，信息可以从许多不同方面提出。更深远的益处包括在模拟模型中使用计算机的能力和使计算机深思的能力。

4.4 网 格 计 算

4.4.1 课文 A

网格架构形成了网格成功应用的核心基础。该架构是一些为有待解决的特定问题、环境而确定的资源和能力的复杂组合。

在交付网格计算应用架构的起始阶段，为了确定特定环境要求的核心架构支持，开发商或服务提供商必须考虑以下一些问题。

- 我们试图为用户解决什么问题？
- 在使用户的应用更便捷的同时，我们如何使网格支持能力简单些？
- 开发商如何帮助用户更快地访问并充分利用这些应用来最大限度地满足解决问题的需求？
- 使用网格工具有多难？
- 网格开发商能为目标用户群体提供灵活可变的环境吗？
- 是否有任何未考虑到的因素，使网格服务提供商更容易为网格创建适合问题域的工具？
- 网格服务提供商应该提出的开放标准、环境和规则是什么？

通常，一个网格计算架构部件必须在实施的多个阶段中处理几个潜在的复杂领域。

- 安全性

在网格计算环境的安全方案中，资源的异构本质及其不同的安全策略是复杂的、复合的。这些计算资源处在不同的安全域和异构平台上。简言之，我们的中间件解决方法必须处理局部安全整合、安全身份映射、安全访问/验证、安全同盟，以及信用管理等问题。其他的安全要求的重点通常在于数据完整性、保密性及信息隐私。

网格计算数据交换必须通过安全通信渠道得以保护，如 SSL/TLS，而且通常与 Web 服务安全性一类的安全信息交换机制相结合。用来保护网格安全的最著名的安全架构是网格安全架构（GSI）。在大多数情况下，网格安全架构使单点登录、异构平台整合及安全资源访问/验证成为可能。最新、最著名的安全解决方案是 Web 服务安全性标准的应用。这种机制为复杂、可操作的安全解决方案提供了信息层面的端到端的安全。

- 资源管理

大量的具有异构可能性的网格计算资源使资源管理成为网格计算环境中重要的、要花大力气来解决的问题。这些资源管理方案常包括资源发现、资源清单、故障隔离、资源供给、资源监控、大量自治能力及服务级管理活动。资源管理区域中最有趣的方面是基于服务级要求的，从网格资源库中甄选出正确的资源并有效地供给以满足用户需求。

多种多样的服务提供商可以管理跨越许多领域的网格计算资源，如安全、管理、网络服务及应用功能等，了解这一点很重要。可操作的和可应用的资源也可以在不同的硬件、软件平台上运行。除这个复杂性外，网格计算中间件必须提供有效的资源监控来收集关于利用情况、可获得性及其他信息的需求矩阵。

我们可以发现另一个有价值且非常关键的网格计算架构的特征，即可以提供自我管理、自我诊断、自我修复、自我设定的自治能力。最著名的资源管理中间件解决方法是网格资源分配管理器（GRAM）。该资源为用户提供了有力的作业管理服务，涵盖了作业分配、状态管理、数据分发，以及开始/终止作业。

- 信息服务

信息服务从根本上关注的是为不同的网格计算架构资源提供有价值的信息。这些服务大大影响并完全依赖信息提供商所提供的信息，如资源的获取、容量和利用。这些信息服务使服务提供商们能够高效地分配与网格计算架构解决方法相关的不同的具体任务所需的资源。此外，开发商及提供商同样可以构建网格解决方法来反映门户网站，并利用元调度器、元资源管理器。这些矩阵有助于服务级管理同资源策略结合。这个信息针对具体资源且是基于从属该资源的图表提供的。我们需要更高级的索引服务或数据整合器及转换器将这些针对具体资源的数据转成对终端用户有价值的信息源。

基于经验，网格计算资源共享的定义已经改变，如今更多的关注放在虚拟组织参与者的协调资源共享的成熟形式上。协调资源共享的应用概念包括虚拟组织中的任何可获得的资源，包括计算能力、数据、硬件、软件及其应用、网络服务和其他任何计算资源获取的形式。

4.4.2 课文 B
地理信息系统（GIS）

GIS，正如美国国家科学基金会定义的，是用于捕获、存储、检索、分析和显示空间（例如位置定义的）数据的计算化数据库管理系统。

一个 GIS 由下列三部分组成：
- GIS 软件。
- 硬件。

运行 GIS 所需的硬件依赖于三个相互相关的可变因素：(1) 作用范围：用途数、应用程序数与用户数；(2) 数据的规模：地图越详细，所需硬件的功能要越强；(3) 功能度：对数据执行的功能或操作的数量与各种功能的复杂性。

- 数据库，分为内部的和外部的两种。

开发 GIS 系统的关键是地理信息编码法。地理信息编码是把属性数据连接到地图的过程。街道地址地理信息编码是商业地理的基础技术。据说，大约 80% 的商业数据有某种类型的地理成分。

地理信息编码看起来简单，但要做好，却要复杂得多。在目录中查找地址在概念上是容易的，但会由于软件、地理信息编码参考目录或数据地址本身中的一些缺点而失败。

GIS 系统允许一系列地图相互重叠。通过观察计算机化地图的组合，零售商能立即看到他的销售额是高还是低，并且他的竞争对手是强还是弱。

GIS 系统允许以各种各样的方式存取数据。大多数全功能 GIS 系统结合了三种基本类型的能力：(1) 表示映射；(2) 使用地图作为组织工具；(3) 空间分析。

GIS 系统有助于其业务开展的领域包括：(1) 房地产；(2) 直销；(3) 保险；(4) 金融；(5) 服务行业；(6) 制造；(7) 运输和销售；(8) 零售。

分析人员说，GIS 市场正被几个因素所驱动：
- 更便宜、更快的硬件。

PC 价格正迅速下跌，而台式计算机正变得功能更强。
- 改进的 GIS 软件。

现今的 GIS 软件更友好，使得没有使用 GIS 经验的用户能迅速学会使用这些系统。
- 降低了 GIS 的价格。

- 种类更多的人口统计数据。
- GIS 更广泛地用于商业。

在市场上存在两类 GIS 应用程序。"开放系统"能从电子表格或数据库程序直接输入数据；"封闭系统"则不能。一般，开放系统更为实用，但对其使用是更有争论的。准备输入 GIS 的属性数据的责任落在了用户肩上。封闭系统是更易于使用的，属性数据整齐地排列在你的面前供你使用。

GIS 设计过程涉及四个基本要素：地理数据、属性数据（内部的与外部的）、地图绘制软件及硬件。关于地理数据，要问的第一个问题是："我对哪个地理区域感兴趣？"属性数据必须与 GIS 兼容，使它能被输入 GIS 中。地图绘制软件应支持数据录入、数据分析、数据输出、数据显示及数据管理。对于硬件，GIS 将很快超过 GIS 销售商所推荐的最小系统需求。努力获取比你目前需要的更多东西。

非空间、内部、外部的属性数据是开发 GIS 的另一个主要成本。购买诸如商业人口统计数据库这样的外部数据和开发内部数据库共占 GIS 系统总成本的 80%。

4.5 电子商务

4.5.1 课文 A

电子商务就是利用电子媒介做生意。它意味着利用简单、快速和低成本的电子通信实现交易，而交易双方无须见面。

从服务的角度来看，电子商务是一种工具，可以满足企业、消费者和管理层在提高商品质量和加快服务交付速度的同时降低服务成本的愿望。

高速的计算机网络使地理上的距离变得微不足道。商人可以销售商品给非传统市场上的客户，开发新的市场及发现商业机会更容易。通过在互联网上全天候地提供产品及服务的最新信息，商家可以与客户、消费者随时建立紧密联系来确保其竞争优势。互联网在网络世界为商家提供了大量的市场和产品推销机会，同时也增强了商家与顾客的关系。利用多媒体技术，可在互联网上有效地建立起企业形象、产品和服务品牌名称。

根据交易的性质对电子商务分类

一般按照电子商务的交易性质分类，可分为以下几个类型。

企业对企业的电子商务（B2B）。目前大多数电子商务都属于这一类，它是组织间的电子市场交易。

企业对个人的电子商务（B2C）。这类电子商务主要是与个体顾客进行的零售交易。亚马逊网站最典型的顾客就是它的消费者或者客户。

消费者对消费者的电子商务（C2C）。这类电子商务指消费者直接向消费者销售产品。例如，个人在分类广告网站发布销售广告，以及销售个人房屋资产、轿车等。个人对个人的电子商务的另一个例子是在互联网上发布个人服务广告和将知识及专家经验作为商品出售。许多拍卖网站也允许个体提供项目来进行拍卖。于是，许多人利用企业的内部网和其他组织的内部网来发布销售或服务的广告。

个人对企业的电子商务（C2B）。这类电子商务包括个人向组织推销产品或服务，以及个人寻找卖家、与卖家互动，并最终达成交易。

互联网电子商务安全

互联网电子商务安全受到来自各方面的威胁。由于来自互联网及黑客截取、滥用、篡改信息内容的威胁，安全漏洞被频繁地讨论。其实，互联网只是不安全性的潜在来源之一，安全问题还有如下一些其他来源。

客户方：在使用或未使用客户设备的情况下，他人可以冒充客户方。使用盗取的信用卡详细资料就是最简单的例子。

卖方：卖方可以进行不适当或不诚实的交易。这类问题包括卖方文档中的客户详细资料被盗取，以及设立网上商店却不提供广告上的商品或服务来骗钱的虚假交易者。

整个网络及通信两端的安全问题可以分为许多种。

1．私密性

当通过电子手段传输信息时，发送者和接收者都希望所传输的信息是保密的，也就是没有被任何第三者看到。因此，除指定的接收者外，传输的信息必须是无法解读的，这样才能赋予电子信息的私密性。

2．身份验证

在系统或用户收到一条电子信息时，发送者的身份必须能够得到验证，以确保发送者身份的可靠性。一般而言，要验证一个用户，至少需要以下各种信息中的一种：你所拥有的东西（如令牌）、你所知道的东西（如个人身份证号码）或者你是什么（如指纹或签名）。

3．完整性

如果信息没有被以任何方式有意或无意地改变，我们就说它保持完整性。对电子商务而言，鉴别购买者发送的订单细节没有被改变是一个主要的安全问题。在电子交易过程中，交易伙伴双方通过电子手段共享的设计规格说明书需要确保没被以任何方式改变，无论是客户发给供应者的，还是供应者发给客户的。

4．不可抵赖性

在涉及应还债务时，因为有人拒绝承认债务的存在，术语"抵赖"就意味着拒绝接受享受的权利和应承担的义务。对电子交易而言，任何一方对交易的单边否认都是令人不可接受的，并有可能导致诉讼案件。设计优良的电子商务系统会提供防止抵赖的方法，也就是提供不可反驳的原始证据、收据和电子信息的内容。从事电子商务的企业常易遭受不可抵赖危险的攻击。

4.5.2　课文 B

做广告就是为了影响买卖双方的交易而进行的信息传播的努力。互联网重新定义了广告的含义。互联网使得消费者能直接与广告商及广告互动。在互动营销中，消费者用鼠标单击广告以得到更多的信息，或发送电子邮件提问题。互联网为广告商提供了双向通信和电子邮件的能力，同时使得广告商把广告费花到他们想针对的特定群体身上，这比传统电话营销更准确。最终，互联网实现了真正的一对一广告。

（1）旗帜广告[①]

旗帜广告是互联网上最常用的广告形式。当你畅游在信息高速公路上时，旗帜广告到处都是。图像的文件大小应该是从 7KB 到 10KB。文件越小，加载速度越快。因为较长的下载时间可能会使浏览者变得没有耐心，或在旗帜广告全部出现之前转移注意力，所以旗帜广告的设计者在图像大小上花费很多心思。通常，旗帜广告包括一篇简短的文字或所推销的产品的图形信息。广告商尽量设计出能吸引消费者注意力的旗帜广告。

随着互联网编程的进步，我们现在开始看到有视频和音频剪辑的旗帜广告。旗帜广告有一些链接，当单击这些链接时，客户就会看到广告商主页。有两种旗帜广告：关键字旗帜广告和随机旗帜广告。当一个预先确定的词在搜索引擎中被查询时，关键字旗帜广告就会出现。对于想缩小目标观众范围的公司而言，这是非常有效的。随机旗帜广告随机出现。想推出自己的新产品的公司要使用随机旗帜广告。

利用旗帜广告的一个主要优势是能够为目标观众定制。人们可以决定针对哪个细分市场。旗帜广告甚至可以定制成一对一的目标广告。"强迫广告"营销策略也被运用，强迫广告就是强迫观众要看的广告。其缺点是整体成本很高。一家公司要获得营销活动的成功，它需要把广告预算中的很大一部分用于获得高的 CPM[②]。

在别人的网站上放置旗帜广告，有几种不同的放置形式。最常见的形式是：交互旗帜广告、交换旗帜广告和付费广告。交互旗帜广告指 A 公司同意展示 B 公司的旗帜广告，以换取 B 公司展示 A 公司的广告。它是一个在网站之间的直接链接。这可能是建立和维持旗帜广告的最便宜的形式，但它却难以安排。交互旗帜广告常常不能实现，因为很难匹配。不过，如果有几家公司参与，多公司的匹配可能会更容易些。交换旗帜广告的组织安排有三个或更多合作伙伴参加交易。付费广告指在互联网上购买旗帜广告空间，类似于在其他媒体上购买广告空间。

（2）弹出窗口

弹出窗口就是初始的网站页面，用来在短时间内捕捉用户的注意力，作为一种促销或向网站主页的引导，或者告诉用户需要什么样的浏览器及其他软件来浏览网页。弹出窗口胜过其他任何广告方法的主要优点是可以创造新颖的多媒体效果，或者为一次访问提供足够多的资料。

（3）场地租赁

搜索引擎往往在其首页提供空间做商业租赁。租期取决于网站主人和承租人的合同协议。不像在不同的时段出现的旗帜广告，场地中的广告位置一直都不变；这样，减少了竞争。场地租赁的劣势是广告往往小而受限制，致使有些浏览者错过了广告。同时，成本也很高。

（4）电子邮件

另一种在网上做广告的方法是购买电子邮件地址，并把该公司的资料寄给在名单上的公司。这种做法的优点是低成本和联系多元化目标受众。大多数公司都建立了一个客户数据库，向其发送电子邮件。电子邮件作为一种营销渠道正在兴起，它提供经济高效的

---

① 旗帜广告也称为横幅广告。
② CPM（Cost Per Million）指每千人成本或千人印象成本。

实施方式和比其他广告渠道更好、更快的响应率。一个电子邮件地址列表可以是一个非常强大的工具，因为针对的是你有所了解的一群人。

4.6 神经网络与专家系统

4.6.1 课文 A

人工神经网络（ANN）是一个基于连接计算方式的信息处理的数学或者计算模型。这项技术的最初灵感来自对大脑中由神经元和神经突触构成的生物电网络的研究。

神经网络的结构

一个神经网络是一组处理单元（节点），这些处理单元由通过它们进行通信的连线连接起来。每个单元都以其随时间变化的激活状态为特征。从当前单元的激活状态出发，计算它发送给该网络的一个信号。这一信号顺着一些连接传到其他一些节点。在传送期间，它会被减弱或加强，这取决于连接的特征。从邻接单元到达该单元的信号被组合成一个输入信号，从该输入信号计算该单元的下一个激活状态。

神经网络由下列要素组成。

- 节点集合
- 连接
- 计算输入信号的一些规则
- 激活函数
- 输出信号函数

典型的前馈神经网络是一个节点集。这些节点中的一些节点被指定为输入节点，另一些节点被指定为输出节点，在这两者之间的是隐藏节点。神经元间的每个连接都有一个用数字表示的权，当神经网络运行时，每个输入节点都会被赋予一个值，这些值由人类操作员从环境传感器或者一些外部程序输入。然后每个节点把给定的值传给从它开始的连接点，在每个连接点上这个值乘以与该节点相关联的权值。下一层的每个节点随后接收到导向自身的节点产生的值的和，而且每个节点上进行针对该数值的简单计算——这个就是典型的 S 型函数。把计算结果传给后续层的节点，重复这个过程直至到达输出节点。

一般来说，神经网络的权值被初始化为一些小的随机数。这意味着网络什么都不知道，它的输出从本质上来说是其输入的一个随机函数。随着训练过程不断进行，连接点的权值根据特定的学习算法的计算规则逐渐改变。理想地，权值最后收敛为一个允许它们进行有效计算所需的值。

神经网络的类型

- 单层感知器

最早的神经网络类型是单层感知器网络，它是由单一的输出节点层组成的，输入通过一个权值序列直接传递给输出。由于采用这种方式，所以它被认为是最简单的前馈神经网络。

- 多层感知器

这类网络由多层的计算单元组成，它们通常以前馈的形式相互连接。每层中的每个神经元与后继层的神经元之间有直接的连接。在许多应用方面，这些网络的单元把 S 型函数作为激活函数。

多层网络使用各种各样的学习技术，其中最流行的是反向传播。这个方法将输出的值和准确答案进行比较，来计算某个预先确定的误差函数的值。通过各种各样的技术，误差经过网络反馈。利用这个信息，算法对每个节点的权值进行调整，把误差函数值减少到某个小的数量级。这些操作重复了相当多个训练周期后，网络通常收敛到某个状态，该状态下的计算误差很小。这样，我们就说这个网络已经学会了某个目标函数。

神经网络的优点

由于类似神经系统结构的原理，ANN 有许多优势。
- 学习能力：ANN 有能力在所谓的学习阶段基础上进行学习。
- 自组织：ANN 在学习过程中创造了对给定数据的自身表示法。
- 容错：由于存储了多余的信息，ANN 的局部破坏不会完全损坏神经网络的反应。
- 灵活性：ANN 可以处理没有重大变化的输入数据，如噪声信号，或给定输入数据的其他变化。
- 实时：ANN 的结构是并行的，如果通过使用计算机或特殊的硬件来执行，实时是可以实现的。

神经网络目标

虽然各种神经网络模型在不同程度上使人想起某个实际的神经网络，然而它们与它是有区别的（除非这些模型是为了精确模拟而构建的）。施加一些限制的目的是简化推理，并使所设计的算法能执行。人工神经网络能执行某些智能操作，如联想（操作）。对这些网络的分析也提供了关于并行处理的一些有趣结论。

神经单元以很长的响应时间为其特征，但由于大规模并行处理，人类能即时执行传统顺序机需要更多时间的任务，如图像识别。对这些处理的调查研究将最终导致设计使用人工神经细胞的并行性的快速算法。并行处理在构造采用神经网络的搜索算法中是特别重要的。

4.6.2 课文 B
专家系统

依赖人类领域专家的知识作为系统的问题求解策略是专家系统的主要特征。专家系统是使用经过编码的知识来解决专门领域中问题的一组程序，这个专门领域通常需要人的专门知识。专家系统的知识从各种专家源获得，并且以适合于系统在其推理过程中使用的形式编码。

专家知识必须从专家或其他专门知识源，如教科书、杂志文章和数据库中获得。这类知识通常需要在医学、地质、系统配置或工程设计等一些特定领域中的许多培训和经验。一旦收集了足够多的一批专家知识，对其必须以某种形式进行编码，将其放进知识库中，然后对其进行测试，并且在系统的整个生命周期中不断地对其进行改进。

专家系统在以下几个重要方面与常规计算机系统不同。

1. 专家系统使用知识而不是数据来控制求解过程。大部分使用的知识实际上是启发式的，而不是算法型的。

2．知识作为一个与控制程序分开的实体被编码和维护。因此，它不与控制程序一起被编译，这就允许增量式地增加和修改知识库，而不需重新编译控制程序。

3．专家系统能够解释一个特定的结论是怎么得到的，以及在咨询过程中为什么需要所要求的信息。

4．专家系统对知识使用符号表示法，并且通过符号计算进行推理，这非常类似于自然语言处理。

5．专家系统经常用元知识进行推理，即它们用有关自己的知识及自己的知识范围和能力来推理。

专家系统的推理应受到公开检验，提供问题求解状态的相关信息及程序做出选择和决定的解释。对医生或工程师等人类专家来说，如果要他或她接受计算机的建议，合理的解释很重要。实际上，很少人类专家愿意听取他人的建议，更何况是一台不能说明理由的机器。

人工智能和专家系统设计的探索性本质要求程序便于原型化、测试和修改。人工智能的编程语言和环境应支持这种反复迭代的开发方法。例如，在完全的产生式系统中，单一规则的改动不会影响全局的句法。可以添加规则，也可以删除规则，无须过多地更改整个程序。专家系统的设计者们通常认为知识库是否便于修改是衡量系统是否成功的一个主要因素。

专家系统的另一特性是启发式问题求解方法的运用。专家系统的设计者发现，非正规的"窍门"和"经验法则"是教科书和课堂正统理论的必要补充。有时，这些规则以可理解的方式扩充了理论知识，更多的时候仅仅是工作中经验性捷径而已。

有趣的是，大多数专家系统针对的都是相对专业性较强、专家级的领域。这些领域通常已被认真研究过，对问题求解策略有明确的定义。而那些依赖于模糊定义的常识性概念的问题则很难用这些方法解决。无论专家系统的前景如何，过高估计这项技术的能力是一个错误。目前的不足包括：

◆ 难以获取问题领域的 深层 知识。
◆ 缺乏健壮性和灵活性。
◆ 无法提供深入的解释。
◆ 验证困难。
◆ 很少从经验中学习。

尽管有这些不足，专家系统仍在大量的重大应用中证明了其价值。

Unit 5　Computer Security

5.1　Network Security

5.1.1　Text A

Hardware, software, and data are the major assets of computer systems, computer security concerns with them. If there is malicious destruction of a hardware device, erasure of a program or data file, or failure of an operating system file manager, it cannot find particular disk file.[1] In addition, an interception—some unauthorized party has gained access to an asset. There are eliciting copying of program or data files, or wiretapping to obtain data in a network. While a loss may be discovered fairly quickly, a silent interceptor may leave no traces by which the interception can be readily detected. An unauthorized party not only accesses but tampers with an asset. Someone might modify the values in a database, alter a program so that it performs an additional computation, or modify data being transmitted electronically. It it even possible for hardware to be modified. Some cases of modification can be detected with simple measures, while other more subtle changes may be almost impossible to detect. The intruder may wish to add spurious transactions to a network communication system, or add records to an existing database, this is fabrication of an unauthorized party.

Encryption techniques

Encryption is the way to solve the data security problem. There are two kinds of encryption techniques—symmetric key encryption and asymmetric key encryption.

For symmetric key encryption, both parties should have a consensus about a secret encryption key. When A wants to send a message to B, A uses the secret key to encrypt the message. After receiving the encrypted message, B uses the same (or derived) secret key to decrypt the message.[2] The advantage of using symmetric key encryption lies in its fast encryption and decryption processes (when compared with asymmetric key encryption at the same security level). The disadvantages are, first, the encryption key must be exchanged between two parties in a secure way before sending secret messages. Secondly, we must use different keys with different parties. For example, if A communicates with B, C, D and E, A should use 4 different keys. Otherwise, B will know what A and C as well as A and D has been talking about. The drawbacks of symmetric key encryption make it unsuitable to be used in the Internet, because it's difficult to find a secure way to exchange the encryption key.

For asymmetric key encryption, there is a pair of keys for each party: a public key and a private key. The public key is freely available to the public, but only the key owner gets hold of the private key. Messages encrypted by a public key can only be decrypted by its corresponding private key, and vice versa. When A sends message to B, A first gets B's public key to encrypt the message and sends it to B. After receiving the message, B uses his private key to decrypt the message. The advantage comes in the public key freely available to the public, hence free from any key exchange problem. The disadvantage is the slow encryption and decryption process. Almost all encryption schemes used in the Internet uses asymmetric key encryption for exchanging the symmetric encryption key, and symmetric encryption for better performance.[3] Asymmetric key cryptography seems to attain secrecy in data transmission, but the authentication problem still exists. Consider the following scenario: when A sends a message to B, A gets B's public key from the Internet - but how can A know the public key obtained actually belongs to B? Digital certificate emerges to solve this problem.

Authentication

Digital certificate is an identity card counterpart in the computer society. When a person wants to get a digital certificate, he generates his own key pair, gives the public key as well as some proof of his identification to the Certificate Authority (CA). CA will check the person's identification to assure the identity of the applicant. If the applicant is really the one "who claims to be", CA will issue a digital certificate, with the applicant's name, e-mail address and the applicant's public key, which is also signed digitally with the CA's private key.[4] When A wants to send B a message, instead of getting B's public key, A now has to get B's digital certificate. A first checks the certificate authority's signature with the CA's public key to make sure it's a trustworthy certificate. Then A obtains B's public key from the certificate, and uses it to encrypt message and sends to B.

Authentication is an important part of everyday life. The lack of strong authentication has inhibited the development of electronic commerce. It is still necessary for contracts, legal documents and official letters to be produced on paper. Strong authentication is then, a key requirement if the Internet is to be used for electronic commerce. Strong authentication is generally based on modern equivalents of the one time pad. For example tokens are used in place of one-time pads and are stored on smart cards or disks.

Key Words

applicant	申请人,申请者
asset	资产,财产
asymmetric	不对称的,非对称的
attain	达到,获得
authentication	认证,证明,鉴定
authority	权威,权力

certificate	证书
consensus	合意，一致
contract	合同，契约
counterpart	副本，对手
cryptography	密码系统，密码术
decryption	解密，译码
destruction	破坏，毁灭
detect	查明，发现，察觉
drawback	缺点，劣势，弊端
elicit	引出，引起
encryption	加密，编密码
erasure	擦掉，删去
fabrication	制造，装配
inhibit	抑制，压抑
interception	拦截，侦听
intruder	入侵者，干扰者
malicious	蓄意的，恶意的
modification	修改，改良，变更
owner	物主，所有人
pad	便签本
readily	容易地，欣然地
secret	秘密，机密
silent	沉默的，无言的，静止的
spurious	伪造的，欺骗的
symmetric	对称的，均衡的
tamper	干预，篡改
trace	跟踪，追踪，查找
trustworthy	可靠的，值得信赖的
unsuitable	不适合的
vice versa	反之亦然
wiretapping	窃听，窃听器，搭线窃听

Notes

[1] If there is malicious destruction of a hardware device, erasure of a program or data file, or failure of an operating system file manager, it cannot find particular disk file.

说明：句中的"If there is … operating system file manager"引导条件状语从句。

译文：假如蓄意破坏硬件设备、抹除程序或数据文件，以及操作系统文件管理器出故障，都会导致不能找到某一磁盘文件。

[2] After receiving the encrypted message, B uses the same (or derived) secret key to

decrypt the message.

说明:"After receiving the encrypted message"是时间状语从句。

译文:B 收到加密的消息后,用相同的(或最初的)密钥将消息解密。

[3] Almost all encryption schemes used in the Internet uses asymmetric key encryption for exchanging the symmetric encryption key, and symmetric encryption for better performance.

说明:"used in the Internet"是定语,修饰主语"encryption schemes";asymmetric key encryption 是指"非对称密钥加密";symmetric key encryption 是指"对称密钥加密"。

译文:在互联网中几乎所有的加密方案都使用非对称密钥加密来替换对称加密密钥和对称加密,以得到更好的性能。

[4] If the applicant is really the one "who claims to be", CA will issue a digital certificate, with the applicant's name, e-mail address and the applicant's public key, which is also signed digitally with the CA's private key.

说明:CA 指"证书授权机构",是可信任的第三方,它保证数字证书的有效性。CA 负责注册、颁发证书,并在证书包含的信息变得无效后删除(收回)证书。句中的"If"引导条件状语从句,"which is also signed…"是非限定性定语从句。

译文:如果申请人确如自己所声称的,证书授权机构将授予带有申请人姓名、电子邮件地址和申请人公钥的数字证书,并且该数字证书由证书授权机构用其私有密钥做了数字签名。

5.1.2 Text B

With the advancement in technology, information systems have enabled organizations to store huge amounts of important data in computer files. These computer files are stored in computers on company premises which have many flaws in security, and if a hacker wants to steal vital information, it can be fairly easily retrieved by individuals outside of the organization. In recent years the field of network security has witnessed several threats that can be severely damaging to many organizations, and if serious enough could potentially drive these companies into bankruptcy. The extent to which most organizations depend on information systems is overwhelming and it will continue to expand substantially into the future. This is why it is very important for organizations to have well developed security procedures. But with dependability comes responsibility, a responsibility of protecting our organization and its employees, customers and suppliers from any damages that might be a threat by outsides. There is no generic technique that will eliminate one hundred percent of the possible threats, however, if the right plan is developed, and maintained, by the organization, much threat to the company can practically be reduced to a minimal.

There exist many prevention techniques that if implemented correctly, can make a substantial difference in deciding whether the organizations of today will become those of tomorrow. Among the preventative techniques that can be implemented include up to date antivirus software, firewalls, increased user authentication and access control policies, monitoring organizational computer usage, data layering, audits, and emergency recovery plan. Although every approach mentioned above is effective in confronting network security, every organization needs to implement a system consisting of several preventative methods.

- Antivirus software: designed to scan, detect, and eliminate computer viruses. It is only effective in eliminating viruses known at the time during which the program was written.
- Firewalls: security systems with specialized software that controls access to the organization's network by the examining every message packet passing between the two networks and rejecting unauthorized messages or access attempts.
- Increased user authentication and access control policies: implementing password policies for every computer within the organization. The passwords should be changed frequently and employees should have access only to the information necessary to perform their work tasks.
- Monitoring organizational computer usage:
- Data layering: organizations need to determine and classify important information (such as company's financial information, employee data, etc.) and establish different layers of security, hence limiting who can get what and where. Only certain employees should have access to this classified information.
- Audit: identifies all of the controls that govern individual information systems. The IT manager needs to understand the organizational layout, security procedures, and employee's roles and responsibilities in order to thoroughly examine the company and its procedures.
- Emergency recovery plan: a plan that needs to be in effect in case of the worst-case scenario. Everyone in the organization needs to be aware of such a plan and the procedures that would be in effect. The policies should be displayed for everyone to see and should be strictly enforced by the managers of the organization.

Key Words

bankruptcy	破产，倒闭
confront	面对，处理
dependability	可靠性
employee	雇员，员工
establish	建立，创立
overwhelming	势不可挡的，压倒一切的，巨大的
premises	（企业的）房屋，建筑，营业场所
responsibility	责任，职责，负责任
scenario	方案，设想
strictly	严格地，完全地
substantially	本质上，实质上，充分地

5.1.3 Exercises

1. Translate the following phrases into English

(1)公开密钥

(2)非对称密钥

(3)访问控制策略

(4)证书授权机构

(5)数据安全问题

2. Translate the following phrases into Chinese

(1)symmetric key encryption

(2)preventative method

(3)encryption technique

(4)private key

(5)emergency recovery plan

3. Identify the following to be True or False according to the text

(1)There exist many prevention techniques.

(2)The extent to which most organizations depend on information systems is overwhelming and it will continue to expand substantially into the future.

(3)Symmetric key encryption and asymmetric key encryption are almost the same.

(4)Asymmetric key encryption is not the way to solve the data security problem.

(5)The advantage of using symmetric key encryption lies in its fast encryption and decryption processes.

4. Reading Comprehension

(1) When a person wants to get a digital certificate, he generates his own key pair, gives the public key as well as some proof of his identification to the _____.

a. CA

b. grocery

c. cybermarket

d. book store

(2) For_____, there is a pair of keys for each party: a public key and a private key.

a. asymmetric key encryption

b. symmetric key encryption

c. firewall

d. digital certificate

(3)_____ is designed to scan, detect, and eliminate computer viruses.

a. Firewalls

b. Data layering

c. Antivirus software

d. Emergency recovery plan

(4)With the advancement in technology, _____ have enabled organizations to store huge amounts of important data in computer files.

a. operating systems

b. data systems

c. memory systems

d. information systems

5.2　Firewalls and Proxies

5.2.1　Text A

When you connect your LAN to the Internet, you are enabling your users to reach and communicate with the outside world. At the same time, however, you are enabling the outside world to reach and interact with your LAN.

The purpose of a network firewall is to provide a shell around the network which will protect the systems connected to the network from various threats.[1] The types of threats a firewall can protect against include:

◆ Unauthorized access to network resources

An intruder may break into a host on the network and gain unauthorized access to files.[2]

◆ Denial of service

An individual outside the network could, for example, send thousands of mail messages to a host on the net in an attempt to fill available disk space or load the network links.

◆ Masquerading

Electronic mail appearing to have originated from one individual could have been forged by another with the intent to embarrass or cause harm.

Basically, a firewall is a standalone process or a set of integrated processes that runs on a router or server to control the flow of networked application traffic passing through it. Typically, firewalls are placed on the entry point to a public network such as the Internet. They could be considered traffic cops. The firewall's role is to ensure that all communication between an organization's network and the Internet conform to the organization's security policies. Primarily these systems are TCP/IP based and, depending on the implementation, can enforce security roadblocks as well as provide administrators with answers to the following questions.

◆ Who's been using my network?

◆ Who failed to enter my network?

◆ When were they using my network?

- Where were they going on my network?
- What were they doing on my network?

A firewall can reduce risks to network systems by filtering out inherently insecure network services. Network File System (NFS) services, for example, could be prevented from being used from outside of a network by blocking all NFS traffic to or from the network. This protects the individual hosts while still allowing the service, which is useful in a LAN environment, on the internal network. One way to avoid the problems associated with network computing would be to completely disconnect an organization's internal network from any other external system. This, of course, is not the preferred method. Instead what is needed is a way to filter access to the network while still allowing users access to the "outside world".

In this configuration, the internal network is separated from external networks by a firewall gateway. A gateway is normally used to perform relay services between two networks. In the case of a firewall gateway, it also provides a filtering service which limits the types of information that can be passed to or from hosts located on the internal network.[3]

In general, there are three types of firewall implementations, some of which can be used together to create a more secure environment. These implementations are: packet filtering, application proxies, and circuit-level or generic-application proxies.

Packet filtering

Consider your network data a neat little package that you have to driver somewhere. This data could be part of an e-mail, file transfer, etc. With packet filtering, you have access to deliver the package yourself. The packet filter acts like a traffic cop; it analyzes where you are going and what you are bringing with you. However, the packet filter does not open the data package, and you still get to drive it to the destination if allowed.

Most commercial routers have some kind of built-in packet filtering capability. However, some routers that are controlled by ISPs may not offer administrators the ability to control the configuration of the router. In those cases, administrators may opt to use a standalone packet filter behind the router.

Application proxy

To understand the application proxy, consider this scenario where you needed to deliver your neat little package of network data.[4] With application-level proxies, the scenario is similar, but now you need to rely on someone else to deliver the package for you. Hence the term proxy illustrates this new scenario. The same rules apply as they do for packet filtering, except that you don't get to deliver your package past the gate. Someone will do it for you, but that agent needs to look inside the package first to confirm its contents. If the agent has permission to deliver the contents of the package for you, he will.

Of curse, security and encryption also come into play, since the proxy must be able to open the "package" to look at it or decode its contents.

Circuit-level or generic-application proxy

As with application-level proxies, you need to rely on someone to deliver your package for you. The difference is that if these circuit-level proxies have access to deliver the package to your requested destination, they will. They don't need to know what is inside. Circuit-level proxies work outside of the application layers of the protocol. These servers allow clients to pass through this centralized service and connect to whatever TCP port the clients specify.

Key Words

conform	使一致，依照
cop	警察
denial	拒绝接受，否认
embarrass	妨碍，阻碍
masquerade	冒充，假装
neat	干净的，匀整的
opt	选择，赞成
prefer	更喜欢
proxy	代理，代表
roadblock	路障
shell	外壳

Notes

[1] The purpose of a network firewall is to provide a shell around the network which will protect the systems connected to the network from various threats.

说明："to provide…"是表语，"around the network"是"shell"的定语，"which"引导的定语从句也修饰"shell"。

译文：网络防火墙的目的是在网络周围设置一层外壳，用于防止连入网络的系统受到各种威胁。

[2] An intruder may break into a host on the network and gain unauthorized access to files.

说明：入侵者可能闯入网上的主机，并对文件进行非授权访问。

译文："An intruder"是本句的主语，"break into…"和"gain unauthorized"是并列的，主语都是"An intruder"。

[3] In the case of a firewall gateway, it also provides a filtering service which limits the types of information that can be passed to or from hosts located on the internal network.

说明：本句中的"it"指的是"firewall gateway"；"which…"是宾语补足语，解释说明"filtering service"的功能。

译文：防火墙网关还提供过滤业务，它可以限制进出内部网主机的信息类型。

[4] To understand the application proxy, consider this scenario where you needed to deliver

your neat little package of network data.

说明:"To understand the application proxy"是目的状语;主句是"consider this scenario";"where"引导的是补语,进一步解释说明"scenario"。

译文:为理解应用程序代理,请考虑这样的情况,你需要递交一个干净的小网络数据包。

5.2.2 Text B

Network security is the protection of hardware, software and the data of system in the network system and the system will run normally without any damage, change and leak. The innate character of network security is the information security in the network. In a broad sense, the technology and theory involved of privacy, integrity, usability and authenticity of information in the network is belong to the study field of network security. Network security is a comprehensive subject involved computer science, network technology, communication technology, cryptography, information security technology, math application, math theory and information theory.

Attack prevention

Different security mechanisms can be used to enforce the security properties defined in a given security policy. Depending on the anticipated attacks, different means have to be applied to satisfy the desired properties. Attack prevention is a class of security mechanisms that contains ways of preventing or defending against certain attacks before they can actually reach and affect the target. An important element in this category is access control, a mechanism which can be applied at different levels such as the operating system, the network, or tile application layer.

Access control limits and regulates the access to critical resources. This is done by identifying or authenticating the party that requests a resource and checking its permissions against the rights specified for the demanded object. It is assumed that an attacker is not legitimately permitted to use the target object and is therefore denied access to the resource. As access is prerequisite for an attack, any possible interference is prevented.

The most common form of access control used in multi-user computer systems are access control lists for resources that are based on the user and group identity of the process that attempts to use them. The identity of a user is determined by an initial authentication process that usually requires a name and a password. The login process retrieves the stored copy of the password corresponding to the user name and compares it with the presented one. When both match, the system grants the user the appropriate user and group credentials. When a resource should be accessed, the system looks up the user and group in the access control list and grants or denies access as appropriate. An example of this kind of access control can be found in the UNIX file system, which provides read, write and execute permissions based on the user and group membership. In this example, attacks against files that a user is not authorized to use are prevented by the access control part of the file system code in the operating system.

A firewall is an important access control system at the network layer. The idea of a firewall is based on the separation of a trusted inside network of computers under single administrative control from a potential hostile outside network. The firewall is a central choke point that allows enforcement of access control for services that may run at the inside or outside. The firewall prevents attacks from the outside against the machines in the inside network by denying connection attempts from unauthorized parties located outside, in addition, a firewall may also be utilized to prevent users behind the firewall from using certain services that are outside.

Attack avoidance

Security mechanisms in this category assume that an intruder may access the desired resource but the information is modified in a way that makes it unusable for the attacker. The information is preprocessed at the sender before it is transmitted over the communication channel and post-processed at the receiver. While the information is transported over the communication channel, it resists attacks by being nearly useless for an intruder. One notable exception is attacks against the availability of the information, as an attacker could still interrupt the message. During the processing step at the receiver, modifications or errors that might have previously occurred can be detected (usually because the information can not be correctly reconstructed). When no modification has taken place, the information at the receiver is identical to the one at the sender before the preprocessing step.

The most important member in this category is cryptography, which is defined as the science of keeping message secure. It allows the sender to transform information into what may seem like a random data stream to an attacker, but can be easily decoded by an authorized receiver.

The original message is called plaintext. The process of converting this message through the application of some transformation rules into a format that hides its substance is called encryption. The corresponding disguised message is denoted as ciphertext, and the operation of turning it back into plaintext is called decryption. It is important to notice that the conversion from plain to ciphertext has to be lossless in order to be able to recover the original message at the receiver under all circumstances.

The transformation rules are described by a cryptographic algorithm. The function of this algorithm is based on two main principles: substitution and transposition. In the case of substitution, each element of the plaintext (e.g., bit, block) is mapped into another element of the used alphabet. Transposition describes the process where elements of the plaintext are rearranged. Most systems involve multiple steps of transposition and substitution to be more resistant against cryptanalysis. Cryptanalysis is the science of breaking the cipher, i.e., discovering the substance of the message behind its disguise.

The most common attack, called known plaintext attack, is executed by obtaining ciphertext together with its corresponding plaintext. The encryption algorithm must be so complex that even if the code breaker is equipped with plenty of such pairs, it is infeasible for here to retrieve the key. An attack is infeasible when the cost of breaking the cipher exceeds the value of the information, or the time it takes to break it exceeds the lifespan of the information itself.

Given pairs of corresponding cipher and plaintext, it is obvious that a simple key guessing algorithm will succeed after some time. The approach of successively trying different key values until the correct one is found is called brute force attack because no information about the algorithm is utilized. In order to be useful, it is a necessary condition for an encryption algorithm that brute force attacks are infeasible.

Key Words

alphabet	字母表，字母系统
anticipate	预料，预测
ciphertext	密文
credential	证明信，凭证
cryptographic	加密的，密码的
decode	译码
denote	表示，意味着
disguise	伪装，借口
encryption	加密
infeasible	不可实行的
innate	内在的，固有的
integrity	完整，完好
intruder	闯入者，侵入者
leak	泄露，漏洞
legitimately	正当地，合理地
lifespan	有效期，生命期
lossless	无损的
plaintext	明文
prerequisite	前提，先决条件

5.2.3 Exercises

1. Translate the following phrases into English

(1)访问控制

(2)攻击预防

(3)随机数据流

(4)网络安全策略

(5)网络服务系统

2. Translate the following phrases into Chinese

(1)brute force attack

(2)cryptographic algorithm

(3)attack avoidance

(4)authentication process

(5)security mechanism

3. Identify the following to be True or False according to the text

(1)The packet filter acts like a traffic cop; it analyzes where you are going and what you are bringing with you.

(2)A firewall can't reduce risks to network systems by filtering out inherently insecure network services.

(3)Different security mechanisms can be used to enforce the security properties defined in a given security policy.

(4)The most common attack, called known plaintext attack, is executed by obtaining ciphertext together with its corresponding plaintext.

(5)The identity of a user is determined by an initial authentication process that only requires a name.

4. Reading Comprehension

(1) Most commercial routers have some kind of _____packet filtering capability.

 a. built-in

 b. outside

 c. outer

 d. external

(2) In general, there are _____ types of firewall implementations, some of which can be used together to create a more secure environment.

 a. two

 b. one

 c. four

 d. three

(3)Cryptanalysis is the science of breaking the cipher, i.e., discovering the substance of the message behind its _____.

 a. cover

 b. truth

 c. disguise

 d. reality

(4) An attack is infeasible when the cost of breaking the cipher exceeds the value of the information, or the time it takes to break it exceeds the _____ of the information itself.

    a. time

    b. number

    c. lifespan

    d. date

5.3 Computer Virus

5.3.1 Text A

A computer virus is a self-duplicating computer program that interferes with a computer's hardware or operating system.[1] Viruses are designed to duplicate or replicate themselves and to avoid detection. Like any other computer program, a virus must be executed for it to function—that is, it must be located in the computer's memory, and the computer must then follow the virus's instructions. These instructions are called the payload of the virus. The payload may disrupt or change data files, display an irrelevant or unwanted message, or cause the operating system to malfunction.

Most viruses place self-replicating codes in other programs, so that when those other programs are executed, even more programs are "infected" with the self-replicating codes. These self-replicating codes, when caused by some event, may do a potentially harmful act to your computer.[2]

Categories of computer viruses

A macro virus infects a set of instructions called a "macro". A macro is essentially a miniature program that usually contains legitimate instructions to automate document and worksheet production. Experts estimate that more than 2000 viruses exist. Nevertheless, most virus damage is caused by fewer than 10 viruses. Of these viruses, macro viruses account for more than 90 percent of virus attacks.

Script viruses are written in script programming languages, such as Visual Basic Script and JavaScript. These script languages can be seen as a special kind of macro language and are even more powerful because most are closely related to the operating system environment.

In the computer field, the Trojan virus is a kind of bad intention program. It can be used to do some bad intention behavior. It does not make a direct damage to your computer mostly but control your computer. The mainly spread way are e-mail and software download. The control terminal can send the Trojan program as an appendix with the e-mail. The system receiver will infect it as soon as he opens the file. Some uncommon websites tie the Trojan program to the install software with the mask of software download. When you run the software, the Trojan program will install automatically as well.

Time bombs and logic bombs are two computer viruses which lurk in your computer system without your knowledge. A time bomb is a computer program that stays in your system undetected until it is triggered by a certain event in time, such as when the computer system clock reaches a certain date. A time bomb is usually carried by a virus or Trojan horse. A logic bomb is a computer program that is triggered by the appearance or disappearance of specific data.

A worm is a program designed to enter a computer system—usually a network—through security "holes". Like a virus, a worm reproduces itself. Unlike a virus, a worm doesn't need to be attached to a document or executable program to reproduce.

The spread of computer virus

Viruses can be disguised as attachments of funny images, greeting cards, or audio and video files. Viruses can spread through download on the Internet. They can be hidden in illicit software and other files or programs you might download. And it will spread when you run the files or programs. Some viruses will spread on the Internet at certain time, that may be one day at certain time. Of course, when you use e-mail on the Internet, you must be careful with the uncertain mails. If you are not sure who emailed the letter, you should not open it, as it may be with a virus.

Viruses can also spread through the USB port when you use USB disk, MP3 and so on. Today, there are many viruses designed for the USB. When you insert the USB device, the antivirus software will not scan the USB device automatically, and the USB device will be the virus source to the computer.[3] So we must be careful when we use the MP3 and other movable stored devices.

The symptoms and the harm of virus infection

Viruses remain free to proliferate only as long as they exist undetected. Accordingly, the most common viruses give off no symptoms of their infection. Anti-virus tools are necessary to identify these infections. However, many viruses are flawed and do provide some tip-offs to their infection. Here are some indications to watch for:
- Format the whole disk or the sector of the disk.
- Changes in the length of programs.
- Changes in the file date or time stamp.
- Longer program load time.
- Slower system operation.
- Reduced memory or disk space.
- Unusual error messages.
- Failed program execution.
- Modify and damage the data of the file.
- Delete important information on the hard disk.

- Disturb the operation by the user and damage the common display of the computer.

Preventive methods

There are two common methods used to detect viruses. First, and by far the most common method of virus detection is using antivirus software. The disadvantage of this detection method is that users are only protected from viruses that predate their last virus definition update. The second method is to use a heuristic algorithm to find viruses based on common behaviors. This method has the ability to detect viruses that antivirus security firms have not yet created a signature for.

Many users install antivirus software that can detect and eliminate known viruses after the computer downloads or runs the executable files. They work by examining the content of the computer's memory and the files stored on fixed or removable drives, and comparing those files against a database of viruses. Users must update their software regularly to patch security holes. Antivirus software also needs to be regularly updated in order to gain knowledge about the latest threats.[4]

Key Words

antivirus	防病毒（软件），抗病毒（软件）
appendix	附录，附加物
attachment	附件，附属物
detection	侦查，检查，检测
disappearance	失踪，消失
disguise	伪装
disturb	干扰，打扰
flaw	缺陷，缺点
heuristic	启发式的，探索法的
illicit	违法的，违禁的
infect	传染，散布病毒
intention	意向，意图
legitimate	合法的，正常的
lurk	潜伏，潜在，隐身
patch	补丁，修补
predate	在日期上先于
preventive	预防措施，预防法
proliferate	繁殖，扩散
replicate	复制，复写
reproduce	复制，模拟，再现
script	脚本
signature	签名，署名

symptom	征兆，症状
tip-off	密告，警告，暗示
trigger	触发，激起
worksheet	工作单，记工单，操作单

Notes

[1] A computer virus is a self-duplicating computer program that interferes with a computer's hardware or operating system.

说明：本句由"that..."引导宾语从句，修饰宾语"computer program"。

译文：计算机病毒是干扰计算机硬件或操作系统、能自我复制的计算机程序。

[2] These self-replicating codes, when caused by some event, may do a potentially harmful act to your computer.

说明：本句的"when caused by some event"是插入语，作为时间状语。

译文：当这些自我复制的代码被一些事件触发时，或许会做出一些对计算机有潜在危害的行为。

[3] When you insert the USB device, the antivirus software will not scan the USB device automatically, and the USB device will be the virus source to the computer.

说明："When you insert the USB device"是时间状语，后跟两个并列的句子。

译文：当你插入 USB 设备时，反病毒软件不能自动扫描 USB 设备，该 USB 设备就会成为你的计算机的病毒源了。

[4] Antivirus software also needs to be regularly updated in order to gain knowledge about the latest threats.

说明：本句的"in order to"引导目的状语从句。

译文：反病毒软件也需要经常更新，以获取最新的威胁信息。

5.3.2　Text B

Computer virus is a set of computer order or program code which written to damage computer functions or data so that to affect the use of computer and can copy by itself. It can make use of the defects of computer hardware and software to infect the inter data of computer. In order to infect a computer, a virus has to have the chance to execute its code. Viruses usually ensure that this happens by behaving like a parasite, i.e. by modifying another item so that the virus code is executed when the legitimate item is run or opened.

How to detect viruses

New viruses are being developed every day. New techniques may render existing preventive measures insufficient. The only truth in the virus and anti-virus field is that there is no absolute security. However, we can minimize the damage by identifying virus infections before they carry out their payload. The following lists some ways to detect virus infections:

- Watch out for any changes in machine behavior. Any of the following signs could be symptoms of virus activity. Programs take longer time than usual to execute or there is a sudden reduction in the available system memory or disk space.
- A memory-resident anti-virus software can be employed to continuously monitor the computer for viruses.
- Scan your hard disk with anti-virus software. You should make sure that an up-to-date virus signature has been applied and you should update the signature at least once a month.
- Employ server-based anti-virus software to protect your network. You should also consider employing application-based anti-virus software to further protect your machine.

How to clean viruses

Virus has been found? Don't panic! Some pieces of advice about removing computer virus are given below:

- All activities on the infected machine should be stopped (and it should be detached from the network) as the payload may be triggered at any time. Continuing the use of the infected machine helps the suspected virus spread further.
- Recover from backup is the most secure and effective way to recover the files.
- In some cases, you may recover the boot sector, partition table and even the BIOS data using the emergency recovery disk.
- In case you do not have the latest backup of your files, you may try to remove the virus using anti-virus software.

Anti-virus measures

The fight against computer viruses involves five kinds of counter-measures:

1) Preparation includes making backups of all software (including operating systems) and making a contingency plan.

2) Prevention includes creating user awareness, implementing hygiene rules, using disk authorization software, or providing isolated "quarantine" PCs.

3) Detection involves the use of anti-virus software to detect, report and (sometimes) disinfect viruses.

4) Containment involves identifying and isolating the infected items.

5) Recovery involves disinfecting or removing infected items, and recovering or replacing corrupted data.

Recovery methods

Once a computer has been compromised by a virus, it is usually unsafe to continue using the same computer without completely reinstalling the operating system. However, there are a number of recovery options that exist after a computer has a virus.

- Data recovery

We can prevent the damage done by viruses by making regular backups of data (and the Operating System) on different media, that are either kept unconnected to the system, or not accessible for other reasons, such as using different file systems. In this way, if data is lost through a virus, one can start again using the backup.

- Virus removal

One possibility is a tool known as system restore, which restores the registry and critical system files to a previous checkpoint. Often, a virus will cause a system to hang, and a subsequent hard reboot will render a system restore point from the same day.

- Operating System reinstallation

As a last ditch effort, if virus is on your system and antivirus software can't clean it, then reinstalling the Operating System may be required. To do this, the hard drive is completely erased and the Operating System is installed from media (e.g. CDROM). Do not forget, important files should first be backed up before you reinstall the Operating System.

Key Words

checkpoint	检查站，检查点
contingency	意外事件，偶发事件
disinfect	清除（病毒），给……消毒
hygiene	卫生，卫生学
insufficient	不足的，不够的
parasite	寄生，寄生物
partition	分割，分区
payload	有效载荷
quarantine	隔离，隔离期
recovery	恢复，还原
registry	注册表
reinstallation	重新安装
removal	除去，移走
render	给予，提供
restore	恢复，修复

5.3.3 Exercises

1. Translate the following phrases into English

(1)计算机病毒
(2)特洛伊木马
(3)病毒感染
(4)应急恢复盘

(5)反病毒软件

2. Translate the following phrases into Chinese

(1)time bomb

(2)logic bomb

(3)script programming languages

(4)self-duplicating

(5)removable drive

3. Identify the following to be True or False according to the text

(1)New viruses are being developed every day.

(2)A macro virus infects a set of instructions called a "macro".

(3)Viruses can't be disguised as attachments of funny images, greeting cards, or audio and video files.

(4)Many users install antivirus software that can detect and eliminate unknown viruses after the computer downloads or runs the executable files.

(5)In order to infect a computer, a virus has to have the chance to execute its code.

4. Reading Comprehension

(1)A macro is essentially a miniature program that usually contains legitimate _____ _____ to automate document and worksheet production.

   a. instructions

   b. messages

   c. files

   d. data files

(2)Some uncommon websites tie the _____ to the install software with the mask of software download.

   a. script viruses

   b. macro viruses

   c. logic bombs

   d. Trojan program

(3)Computer virus is a set of computer order or _____ which written to damage computer functions or data so that to affect the use of computer and can copy by itself.

   a. database

   b. data

   c. program code

   d. file

(4)A memory-resident anti-virus software can be employed to continuously monitor _____ for viruses.

   a. the computer

   b. the program

c. the database
d. the user

5.4 分词、动名词和动词不定式

5.4.1 分词

  分词是非谓语动词的一种。分词有现在分词和过去分词两种。规则动词的现在分词由动词原形加-ing构成，过去分词由动词原形加-ed构成；不规则动词的分词形式，其构成是不规则的。分词没有人称和数的变化，具有形容词和副词的作用；同时还保留着动词的特征，只是在句中不能独立作为谓语。

  现在分词所表示的动作具有主动的意义，而及物动词的过去分词表示的动作具有被动的意义。现在分词与过去分词在时间关系上，前者表示的动作正在进行，后者表示的动作往往已经完成。现在分词表示的动作与谓语动词表示的动作相比，具有同时性，而过去分词则具有先时性。分词在各种时态、语态下的表现形式如下。

时态	与主语动词同时	doing
	先于主语动词	having done
语态	表示主动	现在分词 doing
	表示被动	过去分词 done
	表示动作已经发生	不及物动词的过去分词

  分词在句子中具有形容词词性和副词词性，可以充当句子的定语、表语、状语和补足语。下面分别举例说明现在分词和过去分词在句子中的作用。

1. 现在分词

- 作为定语

例：An atom contains small particles carrying two kinds of electricity.

译文：原子含有带两种电荷的粒子。

- 作为表语

例：The result of the experiment was encouraging.

译文：实验结果令人鼓舞。

- 作为补足语

例：You'd better start the computer running.

译文：你最好启动计算机。

- 作为状语

例：While making an experiment on an electric circuit, they learned of an important electricity law.

译文：他们在做电路实验时，学到了一条重要的电学定律。

2. 过去分词

- 作为定语

例：The heat energy produced is equal to the electrical energy utilized.
译文：产生的热能与所用的电能相等。
- 作为表语

例：Some substances remain practically unchanged when heated.
译文：有几种物质在受热时几乎没有变化。
- 作为补足语

例：I don't know if we can get the computer repaired in time.
译文：我不知道我们能否按时修好计算机。
- 作为状语

例：Given the voltage and current, we can determine the resistance.
译文：已知电压和电流，我们就可以求出电阻。

5.4.2 动名词

动名词是一种非谓语动词，由动词原形加词尾-ing构成，形式上和现在分词相同。由于动名词和现在分词的形成历史、意义和作用都不一样，通常把它们看成两种不同的非谓语动词。它没有人称和数的变化。动名词具有动词词性和名词词性，因而又可以把它称为"动词化的名词"和"名词化的动词"。动名词在句中充当主语、表语、定语和宾语等，动名词也可以有自己的宾语和状语，构成动名词短语。动名词在各种时态、语态下的形式如下。

动名词	时态/语态	主动	被动
	一般式	doing	being done
	完成式	having done	having been done

例：Excuse me for coming late.
译文：对不起，我来晚了。
句中，coming 是动名词，late 是 coming 的状语。
下面分别举例说明动名词在句子中的作用。

1. 作为主语

动名词作为主语表示一件事或一个行为，其谓语动词用第三人称单数。

例：Changing resistance is a method for controlling the flow of the current.
译文：改变电阻是控制电流流动的一种方法。

动名词作为主语时，也可用 it 作为形式主语，放在句首，而将真正的主语——动名词短语放在谓语之后。

例：It's no good using this kind of material.
译文：采用这类材料是毫无用处的。

2. 作为宾语

动名词可以在一些及物动词和介词后作为介词宾语。要求动名词作为宾语的常用及物动词有 finish、enjoy、avoid、stop、need、start、mean 等。

This printer needs repairing.

译文：这台打印机需要修理。

在英语中，suggest、finish、avoid、stop、admit、keep、require、postpone、practice、fancy、deny 等动词都用动名词作为宾语，不能用不定式作为宾语。但是，在 love、like、hate、begin、start、continue、remember、forget、regret 等词后面可以用动名词作为宾语，也可用动词不定式作为宾语。

例：Do you like watching/to watch TV ?

译文：你喜欢看电视吗？

动名词作为宾语时，如本身带有补足语，则常用 it 作为形式宾语，而将真正的宾语——动名词放在补足语的后面。

例：I found it useless arguing with her.

译文：我发现与她辩论没有用。

下例是动名词作为介词的宾语：

例：Thank you for giving me so much help.

译文：谢谢你给我那么多的帮助。

3. 作为表语

动名词作为表语是名词性表语，表示主语的内容，而不说明主语的性质。主语常为具有一定内涵的名词，这一点与不定式作为表语相似。动名词作为表语与进行时的区别在于主语能否执行该词的行为。若能执行，则为进行时；否则，动名词作为表语（系表结构）。

例：The function of a capacitor is storing electricity.

译文：电容器的功能是存储电能。

4. 作为定语

动名词作为定语是名词性定语，说明名词的用途，与所修饰名词之间没有逻辑主谓关系，这一点是与现在分词作为定语相区别的关键。动名词作为定语时只能使用单词，不可用动名词短语；只能放在所修饰名词的前面，不可后置。

例：English is one of the working languages at international meeting.

译文：英语是国际会议上使用的工作语言之一。

5. 作为宾语补足语

动名词在句中的作用相当于名词，故可作为宾语补足语。动名词只能在少数动词后作为宾语补足语，补充说明宾语的性质、行为或状态，与宾语具有逻辑主谓关系。

例：We call this process testing.

译文：我们称这个过程为检测。

句中，动名词 testing 作宾语 this process 的补足语。

5.4.3 动词不定式

动词不定式是非谓语动词的一种，由不定式符号 to 加动词原形构成。之所以称为"不定式"，是因为它的形式不像谓语动词那样受到主语人称和数的限制。但是，动词不定式又具有动词的许多特点：它可以有自己的宾语、状语及宾语补足语。动词不定式和它的宾语、状语及宾语补足语构成不定式短语。动词不定式还有时态和语态的变化，具体信息如下。

时态	语态（主动）	语态（被动）	用法
一般式	to do	to be done	表示动作有时与谓语动词表示的动作同时发生，有时发生在谓语动词的动作之后
进行式	to be doing		表示动作正在进行，或与谓语动词表示的动作同时发生
完成式	to have done	to have been done	表示的动作发生在谓语动词表示的动作之前

例：Today we use computers to help us do most of our work.

译文：如今，我们使用计算机帮我们做大部分工作。

句中，动词不定式 to help 带有宾语 us 和宾语补足语 do most of our work。

动词不定式通常具有名词性、形容词性和副词性，因此可以充当句子的主语、表语、宾语、定语、状语和补足语。下面分别叙述动词不定式在句中的作用。

1. 作为主语

动词不定式（短语）作为主语，较多地用来表示一个特定的行为或事情，谓语动词需要用第三人称单数，且常用 it 作为形式主语。

例：To know something about computer is important.

译文：懂得一些计算机的知识很重要。

句中，To know something about computer 是动词不定式短语，在句子中作为主语。

不定式短语作为主语时，为了句子的平衡，常常把它放在句尾。而用 it 作为形式主语代替不定式放在句首。

例：It is necessary to learn Visual BASIC.

译文：学习 Visual BASIC 是很有必要的。

句中，It 是形式主语，而真正的主语是动词不定式 to learn Visual Basic。

2. 作为表语

不定式可放在系动词后面作为表语。

例：To see is to believe.

译文：眼见为实。

句中，to believe 放在 is 后面作为表语。

3. 作为宾语

不定式（短语）在某些及物动词后可作为宾语。这类及物动词通常有 want、like、wish、hope、begin、decide、forget、ask、learn、help、expect、intend、promise、pledge 等。

例：This helps to save coal and reduce the cost of electricity.

译文：这有助于节约用煤及降低发电成本。

4. 作为定语

动词不定式（短语）作为定语时，通常放在它所修饰的名词（或代词）之后。

例：He never had the chance to learn computer.

译文：他从来没有学习计算机的机会。

句中，to learn computer 是动词不定式，在句中作为定语，修饰和限定 the chance。

5. 作为状语

不定式作为状语可以修饰句中的动词、形容词、副词或全句，主要表示目的、程度、结果、范围、原因等。

例：To meet our production needs, more and more electric power will be generated.

译文：为了满足生产的需要，将生产越来越多的电力。（表示目的）

例：Solar batteries have been used in satellites to produce electricity.

译文：人造卫星上已经用太阳能电池发电。（表示结果）

应该注意，在"too...to"结构的 too 前面有 not、only、but、never 等含有否定意义的词时，后面的不定式就没有否定意义。

例：English is not too difficult to learn.

译文：英语并不难学。

6. 作为宾语补足语

某些及物动词要求不定式作为宾语补足语。宾语补足语是对宾语的补充说明。

例：A force may cause a body to move.

译文：力可以使物体移动。

7. 作为主语补足语

当主动语态的句子变成被动语态时，主动语态句子中的宾语补足语就在被动语态中变成主语补足语。若主动语态中的宾语补足语由动词不定式构成，则该句变为被动语态后它也相应地变为主语补足语。

例：He was asked to do the experiment at once.

译文：有人请他马上做实验。

但是，当 make、let、have、see、hear、watch、notice、feel 等动词的句子变为被动语态时，原来在主动语态时作为宾语补足语的动词不定式这时也变为主语补足语，此时，动词不定式中的 to 不能省略。

例：He was made to finish repairing the printer.

译文：他被迫马上修好打印机。

5.5 习题答案与参考译文

5.5.1 第 5 单元习题答案

5.1 Network Security

1. Translate the following phrases into English

(1)公开密钥　　　　　　　　　　public key

(2)非对称密钥　　　　　　　　　asymmetric key
(3)访问控制策略　　　　　　　　access control policy
(4)证书授权机构　　　　　　　　Certificate Authority
(5)数据安全问题　　　　　　　　data security problem

2. Translate the following phrases into Chinese

(1)symmetric key encryption　　　对称密钥加密
(2)preventative method　　　　　预防方法
(3)encryption technique　　　　　加密技术
(4)private key　　　　　　　　　私有密钥
(5)emergency recovery plan　　　紧急状况恢复计划

3. Identify the following to be True or False according to the text

T　T　F　F　T

4. Reading Comprehension

(1) a　CA
(2) a　asymmetric key encryption
(3) c　Antivirus software
(4) d　information systems

5.2　Firewalls and Proxies

1. Translate the following phrases into English

(1)访问控制　　　　　　　　　　access control
(2)攻击预防　　　　　　　　　　attack prevention
(3)随机数据流　　　　　　　　　random data stream
(4)网络安全策略　　　　　　　　network security policy
(5)网络服务系统　　　　　　　　Network File System

2. Translate the following phrases into Chinese

(1)brute force attack　　　　　　暴力攻击
(2)cryptographic algorithm　　　加密算法
(3)attack avoidance　　　　　　 攻击避免
(4)authentication process　　　　验证过程
(5)security mechanism　　　　　安全机制

3. Identify the following to be True or False according to the text

T　F　T　T　F

4. Reading Comprehension

(1) a　built-in
(2) d　three

(3) c disguise
(4) c lifespan

5.3 Computer Virus

1. Translate the following phrases into English

(1) 计算机病毒　　　　　　　　computer virus
(2) 特洛伊木马　　　　　　　　Trojan horse
(3) 病毒感染　　　　　　　　　virus infection
(4) 应急恢复盘　　　　　　　　emergency recovery disk
(5) 反病毒软件　　　　　　　　anti-virus software

2. Translate the following phrases into Chinese

(1) time bomb　　　　　　　　时间炸弹
(2) logic bomb　　　　　　　　逻辑炸弹
(3) script programming languages　　脚本程序设计语言
(4) self-duplicating　　　　　　自我复制的
(5) removable drive　　　　　　可移动式驱动器

3. Identify the following to be True or False according to the text

T　T　F　F　T

4. Reading Comprehension

(1) a instructions
(2) d Trojan program
(3) c program code
(4) a the computer

5.5.2 第5单元参考译文

5.1 网　络　安　全

5.1.1 课文 A

硬件、软件和数据是计算机系统的主要资源，计算机安全与它们相关。假如蓄意破坏硬件设备、抹除程序或数据文件，以及操作系统文件管理器出故障，都会导致不能找到某一磁盘文件。此外，还有截取——非授权用户访问系统资源。也有程序或数据文件的非法复制，或搭线窃听以获取网络数据。数据丢失可能会很快被发现，但也可能截获者不留下任何容易检查的痕迹。非法用户不仅可以访问计算机资源还可以篡改资源。有人可以修改数据库中的值、更换一个程序，以便完成另外的计算，或修改正在传送的数据。甚至还可能修改硬件。某些情况下可以用简单测量手段检测出所做的修改，但对有些微妙的修改几乎不可能检测出来。入侵者妄图向网络通信系统加入虚假的事务，或向现有的数据库追加记录。这是非法用户的伪造。

加密技术

加密是解决数据安全问题的途径。加密技术有两种——对称密钥加密和非对称密钥加密。

对于对称密钥加密，当事人双方要有一致的密钥。当 A 给 B 要发送消息时，A 用密钥将消息加密，B 收到加密的消息后，用相同的（或最初的）密钥将消息解密。采用对称密钥加密的优点是其加密和解密速度快（与相同安全标准下的非对称密钥加密术相比）。它的缺点是：第一，在发送加密信息之前，当事双方必须安全地交换密钥；第二，对不同当事人，我们必须使用不同的密钥。例如，如果 A 和 B、C、D 及 E 通信，A 必须用四种不同的密钥。否则，B 将知道 A 和 C 及 A 和 D 在谈论什么。要找到安全交换密钥的方式很困难，所以，对称密钥加密的缺点使它不适合用于互联网。

对于非对称密钥加密，当事各方都有一对密钥：一个公钥和一个私钥。可自由使用公钥，但只有密钥持有者拥有私钥。对用公钥加密的消息只能用相应的私钥解密，反之亦然。当 A 给 B 发送消息时，A 首先得到 B 的公钥将消息加密，然后发送给 B。B 收到消息后，用他的私钥将消息解密。这种加密术的优点是人们可以自由获得公钥，因此从交换密钥问题中解脱出来。它的缺点是加密和解密速度慢。在互联网中几乎所有的加密方案都使用非对称密钥加密来替换对称加密密钥和对称加密，以得到更好的性能。非对称密钥加密在数据传输上似乎是安全的，但认证的问题仍然存在。请考虑如下情节：当 A 给 B 发送消息时，A 从互联网上得到 B 的公钥——A 怎样才能知道他获得的公钥确实属于 B？这个问题由数字证书来解决。

认证

数字证书相当于计算机世界的身份证。当一个人想获得数字证书时，他生成自己的一对密钥，把公钥和其他的鉴定证据送达证书授权机构（CA）。证书授权机构将核实这个人的证明，来确定申请人的身份。如果申请人确如自己所声称的，证书授权机构将授予带有申请人姓名、电子邮件地址和申请人公钥的数字证书，并且该数字证书由证书授权机构用其私有密钥做了数字签名。当 A 要给 B 发送消息时，A 必须得到 B 的数字证书，而非 B 的公钥。A 首先核实带有证书授权机构公钥的签名，以确定是否为可信赖的证书。然后，A 从证书上获得 B 的公钥，并利用公钥将消息加密后发送给 B。

认证是日常生活中的重要部分。缺少强有力的认证制约了电子商务的发展。写在纸上的合同、法律文件和官方信函仍是必要的。如果互联网用于电子商务，强有力的认证是一项关键要求。强有力的认证通常建立在现代版的一次性密码本技术上。例如，令牌用来代替昔日的一次性密码本，而且存储在智能卡或磁盘上。

5.1.2 课文 B

随着技术的进步，信息系统已允许组织将大量重要的数据存放在计算机文件中。这些存放在公司计算机中的文件存在许多安全隐患，因为如果黑客想偷窃重要的数据，他可在组织外很容易地取得。近年来，网络安全领域已出现许多严重损害公司利益的威胁，而且如果威胁相当严重则会导致公司破产。大多数公司在很大程度上完全依赖信息系统，而且今后其依赖程度会更高。这也是为什么公司必须启动安全性更好的程序的重要原因。可靠性是保护公司、职员、客户和供应商免于外部威胁的责任。虽然不存在百分之百地消除威胁的一般方法，然而，若能制定和维护正确的计划，公司所受的威胁就能被降到最小。

现在已出现许多预防技术，只要正确加以实施就能确保公司继续发展。可被实现的预防技术包括最新的防毒软件、防火墙、增强用户授权和访问控制策略、监视公司计算机的使用、数据分层、审计和紧急状况恢复计划。虽然以上提到的每种方法在面对网络安全方面都是有效的，但每个公司需要实现一个包含多种预防方法的系统。

- 防毒软件：用于扫描、检测及去除计算机病毒。仅对开发防毒软件时的已知病毒清除是有效的。
- 防火墙：含特殊软件的安全系统，它通过检查在外部网络和公司网络之间通过的每个信息包，以允许外部对公司网络的访问或拒绝未经授权的信息包的访问企图。
- 增强用户授权和访问控制策略：为公司内的每台计算机设置密码。必须经常改变密码，并且职员应只被允许访问其工作必需的信息。
- 监视公司计算机的使用。
- 数据分层：公司必须确定将重要信息（如公司的财务数据、职员数据等）进行分类，并建立不同的安全层次，从而可限制哪些用户及哪里能获取何种信息。只有特定的职员能访问这些分类信息。
- 审计：标明管理各个信息系统的所有控制。IT经理需要明白公司机构、安全流程及职员的角色和职责，从而能彻底审核公司及其流程。
- 紧急状况恢复计划：一个需要在最坏情况下发挥作用的计划。公司中的每个人都应知道这种将发挥作用的计划和流程。政策应能被每个人看到并由公司管理者严格执行。

5.2　防火墙与代理

5.2.1　课文 A

当你把局域网连接到互联网后，你的用户就能够与外部世界进行通信。然而，同时你也让外界能进入局域网并与之交互。

网络防火墙的目的是在网络周围设置一层外壳，用于防止连入网络的系统受到各种威胁。防火墙可以防止的威胁类型包括：

- 非授权的对网络资源的访问

入侵者可能闯入网上的主机，并对文件进行非授权访问。

- 拒绝服务

例如，网络以外的某个人可能向该网上的主机发送成千上万个邮件信息，企图填满可用的磁盘空间，或者使网络链路满负荷。

- 冒充

某个人发出的电子邮件可能被别有用心的人篡改，结果使原发件人感到难堪或受到伤害。

防火墙基本上是一个独立的进程或一组紧密结合的进程，运行在路由器或服务器上，控制经过防火墙的网络应用程序的信息流。一般来说，防火墙置于公共网络（如互联网）

入口处。它们可被看成交通警察。防火墙的作用是确保一个组织的网络与互联网之间所有的通信均符合该组织的安全策略。这些系统基本上是基于 TCP/IP 的,并且取决于实现,它们能实施安全路障并为管理人员提供下列问题的答案。

- 谁一直在使用网络?
- 谁要上网但没有成功?
- 他们在什么时间使用网络?
- 他们在网上去了何处?
- 他们在网上做什么?

防火墙可以通过滤掉某些原有的不安全的网络业务而降低网络系统的风险。例如,网络文件系统(NFS)可以通过封锁进出网络的所有 NFS 业务而防止被网络外部人员利用。这就保护了各个主机,同时使其一直能在内部网中服务,这在局域网环境中很有用。一种回避网络计算相关问题的方法是把组织的内部网与其他任何外部系统完全断开。当然这不是一个好办法,其实需要的是对访问网络进行过滤,同时仍允许用户访问"外部世界"。

在这种配置中,用一个防火墙网关把内部网和外部网分开。网关一般用于实现两个网络之间的中继业务。防火墙网关还提供过滤业务,它可以限制进出内部网主机的信息类型。

通常有三种类型的防火墙实现方案,其中某些可以一起使用,以建立更安全的环境。这些实现方案是包过滤、应用程序代理、电路级或通用应用程序代理。

包过滤

把你的网络数据看成一个你必须送到某个地方的干净的小数据包。该数据可能是电子邮件、文件传输等的一部分。使用包过滤时,你自己来传送该数据包。包过滤器起交通警察的作用,它分析你想到哪里去,你随身携带了什么。但包过滤不打开数据包,如果允许,你仍然要把它送到目的地。

多数商品化的路由器都有某种内建的包过滤功能。然而,有些由 ISP 控制的路由器不可能为管理员提供控制路由器配置的能力。在这些情况下,管理员可能选择使用接在路由器后面的独立包过滤器。

应用程序代理

为理解应用程序代理,请考虑这样的情况,你需要递交一个干净的小网络数据包。使用应用程序级代理,情况是相似的,但现在你需要依靠另外一个人来为你传递该数据包。因此,术语"代理"说明这个新情况。适合包过滤的规则也适合应用程序代理,不同的是,你不能越过应用程序代理来递交包。有人会为你做此事,但该代理人首先要看一下包的内部,确认它的内容。如果代理已有递交该包的内容的许可,他就会为你递交。

当然,由于代理必须有能力打开"包"来看看或者对其内容进行译码,所以安全和加密问题也随之而来。

电路级或通用应用程序代理

与应用程序级代理一样,你需要依靠某个人来为你传递数据包。差别是,如果这些电路级代理要把数据包递交到你要求的目的地,它们就会这么做。它们不需要知道内容是什么。电路级代理工作在协议的应用层的外面。这些服务器允许客户机通过该集中式服务，并连接到客户机指定的 TCP 端口。

5.2.2 课文 B

网络安全是对计算机网络系统中的硬件、软件和系统数据的保护，使系统在没有任何损坏、变化和泄露的情况下正常运行。网络安全的本质是网络中的信息安全。从广义上来说，网络上信息的保密性、完整性、可用性、真实性的相关技术和理论都是网络安全的研究领域。网络安全是一门涉及计算机科学、网络技术、通信技术、密码技术、信息安全技术、数学应用、数学理论及信息论等多种学科的综合性学科。

攻击预防

不同的安全机制可被用来加强既定安全策略限定的安全性能。根据预期的攻击，要使用不同的手段来实现想要的性能。攻击预防是安全机制的一个类别，它包括在攻击真正到来并影响目标之前进行预防或阻止的一些方法。这类方法的一个要素就是访问控制，即可以应用于操作系统、网络或平铺应用层等不同级别的一种机制。

访问控制限定并控制对重要资源的访问。它是通过识别或验证资源请求方并核准其对特定对象的权限来实现的。若攻击者没有合法的许可就使用目标对象则被拒绝访问资源。由于访问是攻击的前提，所以可以防止任何可能的干扰。

在多用户计算机系统中，最常用的访问控制形式是资源访问控制清单。该清单建立在进程中想要使用这些资源的用户和群的身份的基础上。用户身份由最初的验证过程决定，通常要一个用户名和密码。登录过程检索与用户名对应的存储的密码备份并与递交的密码进行比较。当两者都匹配时，系统给予相应的用户和群证书。当一个资源被请求访问时，系统会在访问控制清单中查找用户和群并适当授权访问或拒绝访问。在 UNIX 文件系统中可以找到这种访问控制的例子：文件系统基于用户和群的身份提供读、写、执行的许可。在这个例子中，操作系统中的文件系统代码的访问控制部分阻止了未授权使用文件的用户对文件的攻击。

防火墙是网络层中一个重要的访问控制系统。防火墙的思想是将独立管理控制下的可信任的计算机内部网同有潜在威胁的外部网分开。防火墙是一个主要的阻塞点，能加强对网内或网外服务的访问控制。防火墙通过拒绝来自外部的未授权方的连接企图，从而防止来自外部的对内部网中机器的攻击。除此之外，防火墙还可以用来阻止墙内用户使用外部服务。

攻击避免

这类安全机制认为入侵者可以访问到想要的资源，但可以以某种方式修改信息使攻击者无法使用。信息在通信信道中传送之前由发送者进行了预处理，并且接收者收到信息后也要对信息进行处理。当信息在通信信道传送时，对入侵者来说几乎是无用信息，从而抵制了攻击。一个值得注意的例外是对信息的可用性的攻击，因为攻击者仍然可以中断信息。在接收者处理信息的阶段，已经发生的更改或错误是可以被检查出来的（通常是因为信息不能正确重构）。当没有发生更改时，接收者处理后的信息与发送者进行预处理前的信息是一样的。

这种方法中最重要的成员是密码学，它是一门保持信息安全的学科。它允许发送者将信息转换为在攻击者看来是一种随机数据流的内容，但该内容可以很容易地被授权的接收者解码。

原始信息称为明文。通过应用一些转换规则将信息转换成隐藏其实质内容的格式的过程称为加密。相应的被隐藏了的信息称为密文，将密文转换成明文的过程称为解密。值得注意的是，将明文转换成密文必须是无损的，以便在任何情况下接收者都能恢复原始信息。

转换规则通过密码算法来描述。这种算法的作用基于两个主要的原则：替换和移位。在替换中，明文的每个要素（如字节、块）被映射成所用字母的另一要素。移位是指明文要素重组的过程。大部分系统有多个替换和移位步骤，以便增强对密码分析的抵抗力。密码解析指破解密码的学科，即发现隐藏在伪装背后的信息的实质。

最常见的攻击称为已知明文攻击，是由同时获取密文和相应的明文引起的。加密算法必须足够复杂，以至即使密码破解者同时获得大量密文和明文也无法获取密钥。当破解密码的价值超出信息价值时，或者破解所花时间超出信息本身的使用期限时，攻击则不可行。

如果给定一对相应的密码和明文，很显然一个简单的猜密钥算法在一段时间后总能成功。因为没有使用任何关于算法的信息，这种不断尝试不同密钥值直到找到正确密钥的方法被称为暴力攻击法。有效的加密算法的一个必要条件就是使暴力攻击对其不可行。

5.3 计算机病毒

5.3.1 课文 A

计算机病毒是干扰计算机硬件或操作系统、能自我复制的计算机程序。病毒旨在复制或自我复制并避免被检测到。像任何其他计算机程序一样，病毒必须被执行才能起作用。也就是说，病毒必须在计算机内存里面，而且计算机必须按照病毒的指令运行。这些指令称为病毒的有效载荷。病毒的有效载荷可能扰乱或修改数据文件，显示不相关或不想要的信息，或者导致操作系统发生故障。

许多病毒将可自我复制的代码放置到其他程序中。这样当被感染的程序被执行时，更多的程序就被这些可自我复制的代码感染了。当这些自我复制代码被一些事件触发时，或许会做出一些对计算机有潜在危害的行为。

计算机病毒种类

宏病毒会感染称为"宏"的指令集。宏实质上是一个微型程序，所包含的合法指令通常用来自动完成文档和工作表的操作。专家们认为目前存在 2000 多种病毒，但是，病毒所造成的损失是由不到 10 种病毒引起的。在这些病毒中，宏病毒占病毒攻击的 90%以上。

脚本病毒使用脚本程序设计语言编写，如 Visual Basic 脚本语言和 Java 脚本语言。这些脚本语言可被视为一种特殊的宏语言。它们甚至比宏语言的功能更强大，因为它们大多与操作系统环境有紧密的联系。

在计算机领域中，特洛伊木马病毒是一种用来实现恶意行为的恶意程序。它基本上不会对你的计算机产生直接的破坏，而是对计算机进行控制。主要的传播途径有电子邮件和软件下载。控制端能通过电子邮件的附件发送特洛伊木马程序，系统接收者一旦打开文件，就会被感染。一些不寻常的网站会以软件下载的"面具"将特洛伊木马程序捆绑在安装软件上，当你运行软件时，同时也会自动安装特洛伊木马。

时间炸弹和逻辑炸弹是两种计算机病毒。它们在你没有发觉的情况下潜伏在计算机系统中。时间炸弹是一种计算机程序，它在不被发现的情况下滞留在你的计算机中，直到某个特定的时刻被引发，比如计算机的系统时钟到达了某一天。时间炸弹通常是由病毒或特洛伊木马携带的。逻辑炸弹是一种计算机程序，它是通过某些特定数据的出现或消失触发的。

蠕虫是一种通过安全漏洞来进入计算机系统（通常是网络）的程序。与病毒类似，蠕虫也可以自我复制，但与病毒不同，蠕虫不需要附着在文档或可执行文件上来进行复制。

计算机病毒的传播

病毒以有趣图片、贺卡或者视频和音频文件的附件伪装起来。病毒能通过网上下载的形式进行传播。它们可能隐藏在违法软件及你可能下载的其他文件或程序中。当你运行这些文件或程序时，它就会传播出去。一些病毒能在互联网上在特定时间传播，也许是在某一天的某一时刻。当你在互联网上使用电子邮件时，对于不确定的邮件必须小心谨慎。如果你不确定是谁发给你的邮件，你就不应该打开它，因为它很可能存有病毒。

当你使用 U 盘、MP3 等时，病毒能够通过 USB 接口传播。现今，有很多病毒专门设计用于 USB。当你插入 USB 设备时，反病毒软件不能自动扫描 USB 设备，该 USB 设备就会成为你的计算机的病毒源了。因此，当我们使用 MP3 及其他可移动存储设备时，必须十分小心。

病毒感染的症状及危害

病毒只要在未被发现状态下，就可以自由繁殖。因此，常见的计算机病毒的感染并不显现任何症状。有必要用防病毒工具来识别感染。然而，很多病毒有缺陷，确实常常露出马脚。以下是一些值得留意的迹象：
- 格式化整个磁盘或者磁盘的某个扇区。
- 程序长度变化。
- 文件日期或时间记录变化。
- 程序装入时间加长。
- 系统运行速度降低。
- 内存或磁盘空间减少。
- 异常出错信息。
- 程序运行失败。
- 修改和破坏文件中的数据。
- 删除硬盘上的重要信息。
- 干扰用户的操作及破坏计算机的正常显示。

预防策略

有两种检测病毒的通用策略。第一种方法是迄今为止最常见的病毒检测方法，是使用反病毒软件。这种检测方法的缺点是用户仅仅能够防治那些早于最近更新定义的病毒。第二种方法是使用一种启发式算法来查找基于一般行为的病毒。这种方法能够检测反病毒安全公司尚未创建签名的病毒。

很多用户都安装反病毒软件，在计算机下载或运行可执行文件后可检测和清除已知病毒。它们通过审查计算机内存的内容及存放在固定或可移动驱动器上的文件，将这些文件与病毒库进行比较的方式进行工作。用户必须定期更新软件来修补安全漏洞。对反病毒软件也需要经常更新，以获取最新的威胁信息。

5.3.2 课文 B

计算机病毒是一种破坏计算机功能或数据从而影响计算机使用的一套计算机代码或程序片段，具有自身复制的功能。它通过利用计算机硬件和软件上的缺陷来影响计算机内部的数据。为了要传染某台计算机，病毒必须有机会运行它的代码。病毒传染通常像寄生虫一样，即通过修改其他程序，以便当合法的程序被运行或打开时病毒代码得到运行。

如何检测病毒

每天都有新的病毒被开发出来。新的技术可能使现有的预防措施不足以应对病毒。在病毒与防病毒领域中，唯一的真理就是不存在绝对的安全。然而，我们可通过在病毒感染执行其有效载荷之前识别它们，从而把破坏降低到最低。下面列举了检测病毒感染的一些方法：

- 密切注意机器运转情况的任何变化。下面的任何迹象都可能是病毒活动的征兆：程序执行需要的时间比通常多，或者可用的系统内存或磁盘空间突然减少。
- 可使用常驻内存的防病毒软件不间断地监控计算机病毒。
- 使用防病毒软件扫描你的硬盘。你应该确保应用了最新的病毒特征码，而且应该一个月至少更新一次特征码。
- 运用基于服务器的防病毒软件保护你的网络。你也应该考虑运用基于应用程序的防病毒软件，来进一步保护你的机器。

如何清除病毒

发现了病毒？别惊慌！下面给出消除计算机病毒的一些建议：

- 应该停止受到感染的机器上的一切活动（并应将该机器从网络上断开），因为有效载荷随时可能被触发。继续使用受到感染的机器，会帮助受到怀疑的病毒进一步传播。
- 使用备份进行恢复是恢复文件的最安全、最有效的方法。
- 在有些情况下，可使用应急恢复盘来恢复引导区、分区表及基本输入/输出系统的数据。
- 万一没有最新的文件备份，可以尝试使用防病毒软件消除病毒。

反病毒措施

抗计算机病毒的措施包括五种：

1) 准备措施包括将所有的软件（包括操作系统）制作备份，并制订一个突发预案。

2）预防措施包括提高用户的防范意识，实施安全规划，使用磁盘授权软件，或提供隔离的"隔离"个人计算机。

3）检测措施包括使用反病毒软件进行检测，报告并有时对病毒进行消毒。

4）包容措施包括识别和隔离被传染的程序。

5）恢复措施包括消除或删除被传染的项目，并复原或更换被破坏的数据。

恢复策略

一旦计算机被病毒感染了，在未完全重装操作系统的情况下使用同一台计算机通常是不安全的。然而，在计算机感染病毒后，有一些恢复选项。

- 数据恢复

我们可以通过定期在不同媒介上备份数据（和操作系统）的方法来避免病毒所造成的危害，它们与系统不连接，或者由于其他缘由而不可访问，如使用不同的文件系统。这样，如果由于病毒而使数据丢失，则可以使用备份重新开始。

- 病毒清除

一种可能性是一个被称为系统恢复的工具，它将注册和关键系统文件还原到以前的检查点。通常，病毒会导致系统挂起，而后来的硬启动将提供当天的一个系统恢复点。

- 操作系统重装

作为最后的努力，如果你的系统上有病毒而反病毒软件又不能清除它，则需要重装操作系统。要这样做，必须完全擦除硬盘驱动器，并且从媒介（如只读光盘）来安装操作系统。不要忘记，在重装操作系统之前应先备份重要的文件。

附录 A 计算机专业英语词汇表

A

a user password and code　用户口令和密码
abnormal end　异常终止
abstract data type　抽象数据类型
abstract class　抽象类
access control　访问控制
access permission　访问许可
account　账号
acoustic coupler　声音耦合器
active desktop　活动桌面
active window　活动窗口
acyclic directory structure　非循环目录结构
adapter card　适配卡
adaptive scheduler　自适应调度
address space　地址空间
agent　代理，智能体
algorithm　算法
ambient intelligence　情景智能
animation　动画
antivirus program　防病毒程序
application integration　应用程序集成
application layer　应用层
application object　应用对象
application server　应用服务器
architecture　体系结构
assemble　汇编
asymmetric encryption　非对称加密
asynchronous　异步的
audio-output device　音频输出设备
authentication　鉴别，证实
authorization　授权，认可

B

background 后台
bandwidth 带宽
banner advertising 旗帜广告
bar code reader 条形码读卡器
batch processing 批处理
baud rate 波特率
big data 大数据
binary 二进制
bind 绑定
bitmap 位图
black box 黑盒子
Boolean logic 布尔逻辑
branch 分支
broadcast storm 广播风暴
Browser 浏览器
buffer 缓冲区
bulletin board 告示板，公告板
bus-contention 总线争用

C

carriage return 回车
cellular network environment 蜂窝网络环境
certificate authority 认证中心
channel 信道，通道
chat room 聊天室
check box 复选框
child class 子类
child window 子窗口
chip set 芯片组
cipher text 密文
circuit switching 电路交换
class declaration 类声明
class library 类库
click 单击
click-through ratio 点击率

client 客户机
client/server 客户机/服务器
client-side programs 客户端程序
clipboard 剪贴板
cloud computing 云计算
cloud resources 云资源
cloud services 云服务
command button 命令按钮
comment 注释
communication deadlock 通信死锁
compile 编译
compression 压缩
concurrent control 同期控制
configuration 配置
congestion 拥塞
connectionless service 无连接服务
connection-oriented service 面向连接的服务
console 控制台
content provider 内容提供商
Control Panel 控制面板
cooling fan 冷却风扇
copyright 版权
core dump 内核转储
cracker 黑客
crawler 网络爬虫
critical paths 关键路径
critical region 临界区
cross platform 跨平台的
cryptography 密码学
cybercash 电子现金
cyberspace 网络空间

D

data flow diagram 数据流程图
data link layer 数据链路层
data stream 数据流
data structure 数据结构

database 数据库
database interface 数据库接口
database server 数据库服务器
data window object 数据窗口对象
debugger 调试器
decode 解码
decryption 解密
dedicated line 专用线路
device independent 与设备无关的
diagnosis 诊断
dialog box 对话框
digital camera 数码相机
digital cash 数字现金
digital certificate 数字证书
digital signature 数字签名
digital subscriber 数字用户
digital time-stamp 数字时间戳
direct marketing online 网上直销
directed acyclic graph 有向无环图
diskless workstation 无盘工作站
display adapter 显示适配器
distributed database 分布式数据库
distributed processing 分布式处理
distributed system 分布式系统
downloading 下载
drop-down listbox 下拉式列表框
drop-down menu 下拉式菜单
drum 硒鼓
dynamic binding 动态绑定

E

eavesdropper 窃听者
electronic bank 电子银行
electronic community 电子社区
electronic intermediary 电子中介商
electronic invoice 电子发票
electronic meetings 电子会议

electronic payment system 电子支付系统
electronic procurement 电子采购
electronic wallet 电子钱包
embedded computer 嵌入式计算机
embedded information processing 嵌入式信息处理
embedded real-time system 嵌入式实时系统
emulation 仿真
encapsulation 封装
encoding 编码
encryption 加密
encryption key 加密密钥
end user 终端用户
error correction 纠错
expanded memory 扩充内存
expansion bus 扩展总线
expansion slot 扩展插槽
expert hypermedia 智能超媒体
expert system 专家系统
extended attributes 扩展属性
extended memory 扩展内存
external procedure 外部过程

F

fast packet switching 快速分组交换
fatal error 致命错误
fault tolerance 容错
fiber-optic table 光纤
file server 文件服务器
file system 文件系统
filter 过滤器
Firewall 防火墙
flash memory 闪存
flexible manufacturing systems 柔性制造系统
flow control 流量控制
foreground job 前台作业
foreign agent 外地代理
fragmentation 分段

frame relay　帧中继
front-end　前端，前台
full-duplex　全双工

G

gateway　网关
gateway server　网关服务器
generalized lists　广义表
gigabit network　千兆网
global scheduler　全局调度
graphics tablet　图形输入板
grayscale　灰色标度
grid　网格
grid computing　网格计算
group editor　群编辑器
groupware　群件
guidance　向导，指导

H

Hacker　黑客
hanging indent　悬挂式缩进
hashed file　散列文件
head pointer　头指针
head node　头节点
header and footer　页眉和页脚
heap sort　堆排序
hexadecimal　十六进制
high-level language　高级语言
highlight　高亮度
hits　点击率
holography　全息术
home page　主页
hub　集线器
Huffman codes　霍夫曼编码
hyperlink　超链接
hypermedia　超媒体
hypertext　超文本

hypertext server　超文本服务器

I

icon　图标
identifier　标识符
image　图像
index　索引
indexed file　索引文件
inheritance　继承
insertion sort　插入排序
install　安装
instruction　指令
integrated network　集成网络
intellectual property　知识产权
intelligent interfaces　智能界面
interactive advertisement　交互式广告
interactive video　互动视频
interface　接口，界面
internal sorting　内部排序
internet　互联网
Internet of things　物联网
interrupt　中断

J

job class　作业分类，作业类别
job scheduler　作业调度程序
job object　作业对象
joy stick　控制杆
jurisdiction　权限
justification　对齐，版面调整
just-in-time service　即时服务

K

kernel　核心
key　关键字
keyboard　键盘

keypad　辅助小键盘
keyword　关键字

L

label　标记，记号
laser printer　激光打印机
library　库
line spacing　行间距
linear linked lists　线性链表
linear lists　线性表
link　链接
linked list　链表
linked radix sort　链式基数排序
liquid crystal　液晶
local scheduler　本地调度
local variable　局部变量
log file　日志文件
logical link control　逻辑链路控制
logical schema　逻辑模式
logical structures　逻辑结构
logic circuit　逻辑电路
logic complementation　逻辑补码法
log in　登录
log out　注销登录
low-level language　低级语言

M

machine code　机器码
machine language　机器语言
magnetic pot　磁场
magnetic tape　磁带
mailbomb　邮件炸弹
mailing list　邮件组，邮件清单
main window　主窗口
menu bar　菜单栏
message digest　消息摘要
micro recorder　宏记录器

middleware　中间件
mirror　镜像
mobile commerce　移动商务
mobile communication　移动通信
mobile inventory　移动库存
mobile middleware　移动中间件
mobile user　移动用户
modeling language　建模语言
mother board　主板
multidocument interface　多文档界面
multiline edit box　多行编辑框
multiple inheritance　多重继承
multi-threaded　多线程
multi-processor　多处理器
multitasking　多任务
mutual exclusion　互斥

N

nanosecond　纳秒
narrowband　窄带
natural language　自然语言
navigate　导航
network layer　网络层
network structure　网络结构
network system　网络系统
network administer　网络管理员
node　节点
non-blocking primitive　非阻塞原语
non-impact　非击打式

O

object-oriented　面向对象
object-oriented system　面向对象系统
offline　离线
online　在线
online advertising　在线广告
online business applications　在线商业应用

online entertainment　在线娱乐
online file storage　在线文件存储
online procurement system　在线采购系统
online service　在线服务
online subscription　网上订阅
open system　开放系统
operand　操作数
operational　操作的
optimal scheduling algorithm　最优调度算法
optimal tree　最优树
optional　可选的
ordered tree　有序树
orthogonal list　十字链表
overflow　上溢
overfrequency　超频
Overlapped　重叠
overloading　重载

P

package　包
packet　数据包
packet-filtering gateway　数据包过滤网关
paperless transaction　无纸交易
payment system　支付系统
pull technology　拉式技术
push technology　推式技术
packet switching　包切换
page description language　页面描述语言
page fault　页故障
parallel port　并行接口
payment provider　支付提供商
photo-diode　光敏二极管
photo-resistor　光敏电阻
photo-sensitive drum　感光鼓
pie chart　饼图
pin printer　针式打印机
pixel　像素

platform 平台
plug and play（or PnP, P&P） 即插即用
pointing device 定位设备
point-to-point layer 点对点层
polymorphism 多态
pop-up menu 弹出式菜单
pop-up Window 弹出式窗口
preemptive multitasking 抢先式多任务
private cloud 私有云
private key cryptography 私钥加密
process 进程
programming 编程
protocol 协议
proxy server 代理服务器
public key 公开密钥
public key cryptography 公钥加密

Q

quad speed 四倍速
quadratic probing 二次探测
quantizer 数字转换器，编码器
quantometer 光谱分析仪
query 查询
queue 队列

R

radio button 单选按钮
raster 光栅
real time system 实时系统
receiver 接收者
recipient 收件人
recursive function 递归函数
refresh 刷新
register 注册
repeater 中继器
remote 远程
reset 复位

resident　驻留的
resolution　分辨率
response time　响应时间
routing algorithm　路由算法
response window　响应式窗口
restore　恢复
retrieve　检索
right-click　右击
ring network　环形网络
robotics　机器人技术
router　路由器

S

safe mode　安全模式
scanner　扫描仪
screen capture　屏幕捕获
screen saver　屏幕保护程序
script　脚本
seamless　无缝连接的
search engine　搜索引擎
security certificate　安全认证
serial port　串行端口
service pack　服务包
session　会话
shared variable　共享变量
shopping online　在线购物
shortcut　快捷方式
shortcut key　快捷键
signature　签名
simulator　仿真器
single-line edit box　单行编辑框
smart card　智能卡
social media　社交媒体
source code　源代码
speech recognition　语音识别
speech synthesizer　语音合成器
spreadsheet　电子表格

stock trading online　在线股票交易
structure chart　结构图
subdirectory　子目录
subroutine　子程序
supercomputer　超级计算机
super user　超级用户
switch　交换机
synchronous　同步
system board　系统板，主板
systems administrator　系统管理员

T

taskbar　任务栏
telecommunicating　电信
thread　线程
three-dimensional graphics　三维图像
time-sharing　分时
time slicing　时间分片
toggle switch　拨动开关
touch-sensitive display　触控式显示器
traffic　通信量
traffic control　业务量控制
traffic monitoring systems　交通监控系统
transaction　事务
transaction costs　交易成本
transport layer　传输层
trap　陷阱
trunk cable　中继电缆
trunk line　干线，中继线
typeface　字体
typography　印刷样式

U

ubiquitous computing　普适计算
unauthorized access　未授权访问
undirected graph　无向图
union　共同体

upgrade 升级
upload 上传
upstream 向上传输
Usenet 新闻组
user account 用户账号
user-defined 用户自定义
user ID 用户标识符
user interface 用户界面
user object 用户对象

V

vacuum tube 真空管
vector 矢量
version 版本
video bandwidth 视频带宽
video capture card 视频采集卡
video conferencing system 电视会议系统
video display 视频显示
video game 视频游戏
video telephone 可视电话
video text 可视图文
virtual address space 虚拟地址空间
virtual banking 虚拟银行
virtual community 虚拟社区
virtual circuit packet switching 虚电路分组交换
virtual device 虚拟设备
virtual host service 虚拟主机服务
virtual IP address 虚拟 IP 地址
virtual memory technology 虚拟存储器技术
virtual network 虚拟网络
virus checker 病毒检查程序
visual cues 视觉线索
voice mail 语音邮件
volatile 易失性的
volume label 卷标
voice control 语音控制

W

warm boot　热启动
wave form　波形
wave length　波长
Web page　网页
Web server　Web 服务器
Web site　Web 站点
wheel　Linux 中的特权用户
window painter　窗口画板
windows　窗口
wireless communication　无线通信
wiretapping　窃听
wizard　向导工具
word processing　文字处理
workgroup hub　工作组集线器
workstation　工作站
worm　蠕虫
write protect notch　写保护口

Z

zero access　立即存取
zero complement　补码
zoom in/out　移近/拉远镜头

附录 B 计算机专业英语缩略词

A

AAL（ATM Adaptation Layer） ATM 适配层
ABEOJ（Abnormal End of Job） 作业异常终止
ACL（Access Control Lists） 访问控制表
ACPI（Advanced Configuration and Power Interface） 高级配置和电源接口
ACM（Association for Computing Machinery） 计算机协会
ADSL（Asymmetric Digital Subscriber Line） 非对称用户数字线路
ADT（Abstract Data Type） 抽象数据类型
AGP（Accelerated Graphics Port） 图形加速端口
AI（Artificial Intelligence） 人工智能
AIFF（Audio Image File Format） 声音图像文件格式
ALU（Arithmetic/Logic Unit） 算术/逻辑单元
ANSI（American National Standard Institute） 美国国家标准协会
API（Application Programming Interface） 应用程序设计接口
APPN（Advanced Peer-to-Peer Network） 高级对等网络
ARP（Address Resolution Protocol） 地址分辨/转换协议
ASCII（American Standard Code for Information Interchange） 美国信息交换标准代码
ASP（Application Service Provider） 应用服务提供商
AST（Average Seek Time） 平均寻道时间
ATM（Asynchronous Transfer Mode） 异步传输模式
ATM（Automatic Teller Machine） 自动柜员机
ATR（Automatic Target Recognition） 自动目标识别
AVI（Audio Video Interface） 音频视频接口

B

B2B（Business to Business） 商业机构对商业机构的电子商务
B2C（Business to Consumer） 商业机构对消费者的电子商务
BBS（Bulletin Board System） 电子公告牌系统
BER（Bit Error Rate） 误码率
BFS（Breadth First Search） 广度优先搜索法

BGP（Border Gateway Protocol） 边界网关协议
BIOS（Basic Input/Output System） 基本输入/输出系统
BISDN（Broadband-Integrated Services Digital Network） 宽带综合业务数字网
BLU（Basic Link Unit） 基本链路单元
BOF（Beginning Of File） 文件开头
BPS（Bits Per Second） 每秒的比特数
BRI（Basic Rate Interface） 基本速率接口
BSS（Broadband Switching System） 宽带交换系统

C

C2C（Consumers to Consumers） 消费者对消费者的电子商务
CA（Certificate Authorities） 认证中心
CAD（Computer Aided Design） 计算机辅助设计
CAE（Computer-Aided Engineering） 计算机辅助工程
CAI（Computer Aided Instruction） 计算机辅助教学
CAM（Computer Aided Manufacturing） 计算机辅助制造
CASE（Computer Assisted Software Engineering） 计算机辅助软件工程
CAT（Computer Aided Test） 计算机辅助测试
CB（Control Bus） 控制总线
CCS（Common Channel Signaling） 公共信道信令
CDFS（Compact Disk File System） 紧凑磁盘文件系统
CD-MO（Compact Disc-Magneto Optical） 磁光式光盘
CD-ROM（Compact Disc Read-Only Memory） 只读光盘
CGA（Color Graphics Adapter） 彩色图形适配器
CGI（Common Gateway Interface） 公共网关接口
COM（Component Object Model） 组件对象模型
CPU（Central Processing Unit） 中央处理器
CRC（Cyclical Redundancy Check） 循环冗余校验码
CRM（Client Relation Management） 客户关系管理
CRT（Cathode-Ray Tube） 阴极射线管，显示器
CSMA（Carrier Sense Multi-Access） 载波侦听多路访问
CU（Control Unit） 控制单元
CUI（Command User Interface） 命令用户界面

D

DAC（Digital to Analogue Converter） 数模转换器
DAO（Data Access Object） 数据访问对象

DAP（Directory Access Protocol）目录访问协议
DBMS（Database Management System） 数据库管理系统
DCE（Data Communication Equipment） 数据通信设备
DCE（Distributed Computing Environment） 分布式计算环境
DCL（Data Control Language） 数据控制语言
DCOM（Distributed COM） 分布式组件对象模型
DDB（Distributed DataBase） 分布式数据库
DDE（Dynamic Data Exchange） 动态数据交换
DDI（Device Driver Interface） 设备驱动程序接口
DDK（Driver Development Kit） 驱动程序开发工具包
DDL（Data Definition Language） 数据定义语言
DDN（Data Digital Network） 数据数字网
DEC（Digital Equipment Corporation） 数字设备公司
DES（Data Encryption Standard） 数据加密标准
DFS（Depth First Search） 深度优先搜索法
DFS（Distributed File System） 分布式文件系统
DLL（Dynamic Link Library） 动态链接库
DMA（Direct Memory Access） 直接内存访问
DML（Data Manipulation Language） 数据操纵语言
DMSP（Distributed Mail System Protocol） 分布式电子邮件系统协议
DNS（Domain Name System） 域名系统
DNS（Domain Name Server） 域名服务器
DOM（Document Object Mode） 文档对象模型
DOS（Disk Operation System） 磁盘操作系统
DSM（Distributed Shared Memory） 分布式共享内存
DSP（Digital Signal Processing） 数字信号处理
DVI（Digital Video Interactive） 数字视频交互
DVI（Digital Visual Interface） 数字视频接口

E

EC（Embedded Controller） 嵌入式控制器
EDIF（Electronic Data Interchange Format） 电子数据交换格式
EEPROM（Erasable and Electrically Programmable ROM） 带电可擦除可编程只读存储器
EFT（Electronic Fund Transfers） 电子转账系统
EGA（Enhanced Graphics Adapter） 增强型图形适配器
EGP（External Gateway Protocol） 外部网关协议

EISA（Extended Industry Standard Architecture） 扩展式工业标准结构
EMS（Expanded Memory Specification） 扩充存储器规范
EPH（Electronic Payment Handler） 电子支付处理系统
EPROM（Erasable Programmable ROM） 可擦除可编程只读存储器
ERP（Enterprise Resource Planning） 企业资源计划

F

FAT（File Allocation Table） 文件分配表
FCB（File Control Block） 文件控制块
FCFS（First Come First Service） 先到先服务
FDDI（Fiber-optic Data Distribution Interface） 光纤数据分布接口
FDM（Frequency-Division Multiplexing） 频分多路复用
FDMA（Frequency Division Multiple Address） 频分多址
FDX（Full Duplex） 全双工
FEK（File Encryption Key） 文件加密密钥
FEP（Front Effect Processor） 前端处理机
FIFO（First In First Out） 先进先出
FRC（Frame Rate Control） 帧频控制
FTAM（File Transfer Access and Management） 文件传输访问和管理
FTP（File Transfer Protocol） 文件传输协议

G

GAL（General Array Logic） 通用逻辑阵列
GB（Gigabyte） 千兆字节，吉字节
GDI（Graphics Device Interface） 图形设备接口
GIF（Graphics Interchange Format） 图形转换格式
GIS（Geographic Information System） 地理信息系统
GPI（Graphical Programming Interface） 图形编程接口
GPIB（General Purpose Interface Bus） 通用接口总线
GPS（Global Positioning System） 全球定位系统
GUI（Graphical User Interface） 图形用户接口

H

HDC（Hard Disk Control） 硬盘控制
HDD（Hard Disk Drive） 硬盘驱动器
HDLC（High-level Data Link Control） 高级数据链路控制

HPFS（High Performance File System） 高性能文件系统
HPSB（High Performance Serial Bus） 高性能串行总线
HTML（Hyper Text Markup Language） 超文本标记语言
HTTP（Hyper Text Transport Protocol） 超文本传输协议

I

IAC（Inter-Application Communications） 应用间通信
IAP（Internet Access Provider） 互联网接入服务提供商
IC（Integrated Circuit） 集成电路
ICMP（Internet Control Message Protocol） 互联网控制消息协议
ICP（Internet Content Provider）互联网内容服务提供商
IDC（International Development Center） 国际开发中心
IDE（Integrated Development Environment） 集成开发环境
IDL（Interface Definition Language） 接口定义语言
IEEE（Institute of Electrical and Electronics Engineers） 电气与电子工程师协会
IIS（Internet Information Service） 互联网信息服务
IMAP（Internet Message Access Protocol） 互联网信息访问协议
IP（Internet Protocol） 互联网协议
IPC（Inter-Process Communication） 进程间通信
IRP（I/O Request Packets） 输入/输出请求包
ISDN（Integrated Service Digital Network） 综合业务数字网
ISO（International Standard Organization） 国际标准化组织
ISP（Internet Service Provider） 互联网服务提供商
IT（Information Technology） 信息技术
ITU（International Telecom Union） 国际电信联盟
ITV（Interactive TV） 交互式电视

J

JDBC（Java Database Connectivity） Java 数据库连接
JDK（Java Developer's Kit） Java 开发工具包
JIT（Just In Time） 即时
JVM（Java Virtual Machine） Java 虚拟机

K

KB（Kilobyte） 千字节
KBPS（Kilobits Per Second） 千比特每秒

KMS（Knowledge Management System） 知识管理系统

L

LAN（Local Area Network） 局域网
LAT（Local Area Transport） 本地传输
LBA（Logical Block Addressing） 逻辑块寻址
LCD（Liquid Crystal Display） 液晶显示器
LED（Light Emitting Diode） 发光二极管
LIFO（Last In First Out） 后进先出
LLC（Logical Link Control sub-layer） 逻辑链路控制子层
LP（Linear Programming） 线性规划
LPC（Local Procedure Call） 本地过程调用
LSIC（Large Scale Integration Circuit） 大规模集成电路

M

MAC（Medium Access Control） 介质访问控制
MAN（Metropolitan Area Network） 城域网
MB（Megabytes） 兆字节
MC（Memory Card） 存储卡
MCA（Micro Channel Architecture） 微通道结构
MDA（Monochrome Display Adaptor） 单色显示适配器
MFM（Modified Frequency Modulation） 改进型调频制
MIDI（Musical Instrument Digital Interface） 乐器数字接口
MIMD（Multiple Instruction Stream，Multiple Data Stream） 多指令流，多数据流
MIPS（Million Instructions Per Second） 百万条指令每秒
MIS（Management Information System） 管理信息系统
MISD（Multiple Instruction Stream，Single Data Stream） 多指令流，单数据流
MMU（Memory Management Unit） 内存管理单元
MPEG（Moving Picture Expert Group） 运动图像专家组
MPS（Micro Processor Series） 微处理器系列
MTBF（Mean Time Between Failure） 平均故障间隔时间
MUD（Multiple User Dimension） 多用户空间

N

NAOC（No-Account Over Clock） 无效超频
NAP（Network Access Point） 网络访问点

NCSC（National Computer Security Center） 国家计算机安全中心
NDIS（Network Device Interface Specification） 网络设备接口规范
NFS（Network File System） 网络文件系统
NIS（Network Information Services） 网络信息服务
NOC（Network Operation Center） 网络操作中心
NRU（Not Recently Used） 非最近使用
NSP（Name Server Protocol） 名字服务器协议
NTP（Network Time Protocol） 网络时间协议
NUI（Network User Identifier） 网络用户标识符

O

OA（Office Automation） 办公自动化
OCR（Optical Character Recognition） 光学字符识别
ODBC（Open Database Connectivity） 开放式数据库连接
ODI（Open Data-link Interface） 开放式数据链路接口
OEM（Original Equipment Manufactures） 原始设备制造厂商
OLE（Object Linking and Embedding） 对象链接与嵌入
OOM（Object Oriented Method） 面向对象方法
OOP（Object Oriented Programming） 面向对象程序设计
OS（Operating System） 操作系统
OSI（Open System Interconnect） 开放式系统互连

P

PBX（Private Branch Exchange） 用户级交换机
PCB（Process Control Block） 进程控制块
PCI（Peripheral Component Interconnect） 外部部件互连
PDA（Personal Digital Assistant） 个人数字助理
PDN（Public Data Network） 公用数据网
PEM（Privacy Enhanced Mail） 隐私增强邮件
PHP（Personal Home Page） 个人主页
PIB（Programmable Input Buffer） 可编程输入缓冲区
PIN（Personal Identification Number） 个人身份识别码
POP（Post Office Protocol） 邮局协议
POST（Power-On Self-Test） 加电自检
PPSN（Public Packed-Switched Network） 公用分组交换网
PROM（Programmable ROM） 可编程只读存储器
PSN（Processor Serial Number） 处理器序列号

Q

QC（Quality Control） 质量控制
QLP（Query Language Processor） 查询语言处理器
QoS（Quality of Service） 服务质量

R

RAD（Rapid Application Development） 快速应用开发
RAI（Remote Application Interface） 远程应用程序界面
RAID（Redundant Array Independent Disk） 冗余阵列磁盘
RAM（Random Access Memory） 随机存取存储器
RAM（Real Address Mode） 实地址模式
RARP（Reverse Address Resolution Protocol） 反向地址解析协议
RAS（Remote Access System） 远程访问服务系统
RDA（Remote Data Access） 远程数据访问
RIP（Routing Information Protocol） 路由选择信息协议
RISC（Reduced Instruction Set Computer） 精简指令集计算机
ROM（Read Only Memory） 只读存储器
RPC（Remote Procedure Call） 远程过程调用
RTS（Request To Send） 请求发送

S

SAA（System Application Architecture） 系统应用结构
SACL（System Access Control List） 系统访问控制表
SAF（Store And Forward） 存储转发
SAP（Service Access Point） 服务访问点
SCSI（Small Computer System Interface） 小型计算机系统接口
SDLC（Synchronous Data Link Control） 同步数据链路控制
SDK（Software Development Kit） 软件开发工具箱
SET（Secure Electronic Transaction） 安全电子交易协议
SGML（Standard Generalized Markup Language） 标准通用标记语言
SHTTP（Secure Hype Text Transfer Protocol） 安全超文本传输协议
SIMD（Single Instruction Stream，Multiple Data Stream） 单指令流，多数据流
SISD（Single Instruction Stream，Single Data Stream） 单指令流，单数据流
SMB（Server Message Block） 服务器消息块
SMDS（Switched Multi-megabit Data Services） 交换式多兆位数据服务
SMP（Symmetric Multi-Processor） 对称式多处理器

SMTP（Simple Mail Transfer Protocol） 简单邮件传输协议
SNA（System Network Architecture） 系统网络结构
SNMP（Simple Network Management Protocol） 简单网络管理协议
SNTP（Simple Network Time Protocol） 简单网络时间协议
SONET（Synchronous Optic Network） 同步光纤网
SPC（Stored-Program Control） 存储程序控制
SQL（Structured Query Language） 结构化查询语言
SSIC（Small Scale Integration Circuit） 小规模集成电路
STDM（Synchronous Time Division Multiplexing） 同步时分多路复用
STP（Shielded Twisted-Pair） 屏蔽双绞线
SVGA（Supper Video Graphics Array） 超级视频图形阵列

T

TCB（Transmission Control Block） 传输控制块
TCP（Transmission Control Protocol） 传输控制协议
TCP/IP（Transmission Control Protocol/Internet Protocol） 传输控制协议/网际协议
TDM（Time Division Multiplexing） 时分多路复用
TDMA（Time Division Multiplexing Address） 时分多路复用地址
TFTP（Trivial File Transfer Protocol） 普通文件传输协议
TIFF（Tag Image File Format） 标记图形文件格式
TIG（Task Interaction Graph） 任务交互图
TLI（Transport Layer Interface） 传输层接口
TM（Traffic Management） 业务量管理，流量管理
TTL（Transistor-Transistor Logic） 晶体管—晶体管逻辑电路
TWX（Teletypewriter Exchange） 电传打字机交换

U

UART（Universal Asynchronous Receiver Transmitter） 通用异步收发器
UDF（Universal Disk Format） 通用磁盘格式
UDP（User Datagram Protocol） 用户数据报协议
UI（User Interface） 用户界面，用户接口
UIMS（User Interface Management System） 用户接口管理程序
UNI（User Network Interface） 用户网络接口
UPS（Uninterruptible Power Supply） 不间断电源
URI（Uniform Resource Identifier） 统一资源标识符
URL（Uniform Resource Locator） 统一资源定位器
USB（Universal Serial Bus） 通用串行总线

UTP（Unshielded Twisted-Pair） 非屏蔽双绞线
UXGA（Ultra Extended Graphics Array） 极速扩展图形阵列

V

VAD（Virtual Address Descriptors） 虚拟地址描述符
VAN（Value Added Network） 增值网络
VAP（Value-Added Process） 增值处理
VAS（Value-Added Server） 增值服务
VCPI（Virtual Control Program Interface） 虚拟控制程序接口
VCR（Video Cassette Recorder） 盒式录像机
VDD（Virtual Device Drivers） 虚拟设备驱动程序
VDR（Video Disc Recorder） 光盘录像机
VDT（Video Display Terminals） 视频显示终端
VFS（Virtual File System） 虚拟文件系统
VGA（Video Graphics Array） 视频图形阵列
VGA（Video Graphics Adapter） 视频图形适配器
VHF（Very High Frequency） 甚高频
VIS（Video Information System） 视频信息系统
VLAN（Virtual LAN） 虚拟局域网
VLSI（Very Large Scale Integration） 超大规模集成
VOD（Video On Demand） 视频点播系统
VPN（Virtual Private Network） 虚拟专用网
VR（Virtual Reality） 虚拟现实
VRML（Virtual Reality Modeling Language） 虚拟现实建模语言
VRUs（Voice Response Units） 语音应答系统
VTP（Virtual Terminal Protocol） 虚拟终端协议

W

WAN（Wide Area Network） 广域网
WAE（Wireless Application Environment） 无线应用环境
WAIS（Wide Area Information Service） 广域信息服务，数据库检索工具
WAP（Wireless Application Protocol） 无线应用协议
WAV（Wave Audio Format） 波形音频格式
WDM（Wavelength Division Multiplexing） 波分多路复用
WDP（Wireless Datagram Protocol） 无线数据报协议
WFW（Windows for Workgroups） 工作组窗口
WML（Wireless Markup Language） 无线标记语言

WORM (Write Once, Read Many time) 写一次读多次光盘
WWW (World Wide Web) 万维网
WYSIWYG (What You See Is What You Get) 所见即所得

X

XGA (eXtended Graphics Array) 扩展图形阵列
XML (eXtensible Markup Language) 可扩展标记语言
XMS (eXtended Memory Specification) 扩展存储器规范
XQL (eXtensible Query Language) 可扩展查询语言

参 考 文 献

[1] 吕云翔. 计算机英语实用教程[M]. 北京：清华大学出版社，2015.
[2] 王小刚. 计算机专业英语[M]. 4 版. 北京：机械工业出版社，2015.
[3] 金志权，等. 计算机专业英语教程[M]. 6 版. 北京：电子工业出版社，2015.
[4] 卜艳萍. 计算机专业英语[M]. 4 版. 北京：电子工业出版社，2013.
[5] 王春生，等. 新编计算机英语[M]. 2 版. 北京：机械工业出版社，2013.
[6] 赵桂钦. 电子与通信工程专业英语[M]. 北京：清华大学出版社，2012.
[7] 李心广. 新编计算机英语教程[M]. 北京：人民邮电出版社，2010.
[8] 卜艳萍. 计算机专业英语[M]. 3 版. 北京：电子工业出版社，2009.
[9] 刘兆毓. 计算机英语[M]. 4 版. 北京：清华大学出版社，2009.
[10] 许春勤. 计算机专业英语[M]. 北京：电子工业出版社，2008.
[11] 顾大权. 实用计算机专业英语[M]. 北京：国防工业出版社，2007.
[12] 武马群. 计算机专业英语[M]. 北京：北京工业大学出版社，2005.
[13] http://en.wikipedia.org/wiki/Steve_Jobs.
[14] http://en.wikipedia.org/wiki/Bill_gates.
[15] http://www.thefreedictionary.com/.